Encountering Mobile Data Dynamics in Heterogeneous Wireless Networks

Jie Wang • Wenye Wang • Xiaogang Wang

Encountering Mobile Data Dynamics in Heterogeneous Wireless Networks

Springer

Jie Wang
School of Software Engineering
Tongji University
Shanghai, China

Wenye Wang
Department of Electrical and Computer Engineering
North Carolina State University
Raleigh, NC, USA

Xiaogang Wang
Department of Electrical and Computer Engineering
North Carolina State University
Raleigh, NC, USA

ISBN 978-3-031-62905-1 ISBN 978-3-031-62906-8 (eBook)
https://doi.org/10.1007/978-3-031-62906-8

This Springer imprint is published by the registered company Springer Nature Switzerland AG
The registered company address is: Gewerbestrasse 11, 6330 Cham, Switzerland

Preface

Owing to advances in wireless communication, networking, and data analysis technologies, the generation, dissemination, and acquisition of data are more frequent and accessible than ever. Consequently, data services, which *move* data from its generator(s) to its consumer(s) through individual connections, are quickly migrating to the edge of networks. Such wireless systems are composed of numerous devices that are different in many aspects, e.g., communication technology, mobility pattern, and so on. Though service latency can be greatly reduced at the network edge, emerging data services at the edge create new challenges to the network: heterogeneity of individuals adds to the complexity of system design, management, and performance evaluation, while proliferating end devices impose an ever-increasing demand on resources, that are already scarce at the edge. To exploit the full potential of data services, it is essential to understand the cause, governing rules, and impact of data's *mobility* in such heterogeneous wireless networks, for the benefit of data owners, service providers, and network operators.

Therefore, this book is dedicated to study *mobile data dynamics*, that is, dynamic processes of mobile data. Specifically, we identify three dynamic processes, of *information*, *coverage*, and *spectrum*, as the cause, manifestation, and impact of mobile data, respectively. Then we examine these dynamic processes and the governing rule of data movements, through a modeling and analysis approach to answer the following questions: *When* data move and stop? *Where* data are? *How* data move? *What* impact mobile data induce on network resources?

In particular, we first study conflicting information propagation with a novel *Susceptible-Infected-Cured (SIC)* model to answer the *when* question. Our results reveal the impact of network topology on the lifetime of the undesired information, which provides bounds, scaling laws, and guidelines for practical information control measures. For the *where* question, we quantify the whereabouts of data, that is, data coverage, with a data-strength metric, and find that the change of data coverage depends heavily on user mobility, by which prediction is possible. Then, we consider dissemination processes of multiple data blocks in the emerging DSA-enabled fog paradigm, to answer the *how* and *what* questions. We propose a *gravity model* to describe how data move in an offloading process, based on which we find the amount of storage and communication resource needed for data offloading scales linearly with the network size. Particularly for the spectrum resource, a scarce resource in wireless networks, we study *spectrum activity surveillance (SAS)* to observe the impact of mobile data, and propose multi-monitor deployment strategies with guaranteed performances for both the dedicated and crowdsourcing SAS scenarios. Finally, we recapitulate the key findings with regard to mobile data dynamics, and outline the open research questions in the context of Edge Intelligence (EI). We hope this book helps advance understanding on mobile data and provides design guidelines for future data services in heterogeneous wireless networks.

Shanghai, China — Jie Wang
Raleigh, NC, USA — Wenye Wang
Triangle Park, NC, USA — Cliff Wang
March 2024

Acknowledgment

We would like to express our sincere thanks to all those who worked with us on this book, including the co-authors of our previously published papers that parts of this book are based on, as well as students who helped with the proof-reading. This work is supported by the National Natural Science Foundation of China (Project No. 62102288), the Shanghai Science and Technology Innovation Action Plan (Project No. 23ZR1467300), and the Fundamental Research Funds for the Central Universities. We would also like to express our gratitude to all the editors, reviewers, and staff, who worked on the publication of this book.

Contents

Acronyms

AI	Artificial Intelligence
AP	Access Point
AR	Augmented Reality
BS	Base Station
CDF	Cumulative (probability) Density Function
CDN	Content Delivery Network
CR	Cognitive Radio
D2D	Device-to-Device Communication
DSA	Dynamic Spectrum Access
EI	Edge Intelligence
ER	Erdös-Rényi random graph
ETC	Electric Toll Collecting
GFT	Graph Fourier Transform
GSP	Graph Signal Processing
IoT	Internet-of-Things
ITS	Intelligent Transportation System
LAN	Local Area Network
LBS	Location-Based Service
LTE	Long-Term Evolution
MACC	Mobile Ad hoc Cloud Computing
MCC	Mobile Cloud Computing
MEC	Mobile Edge Computing
MFA	Mean-Field Approximation
MGF	Moment Generating Function
OBU	On-Board Unit
OSN	Online Social Network
PDF	Probability Density Function
PMF	Probability Mass Function
PU	Primary User
QoS	Quality-of-Service
RAT	Radio Access Technology

REM	Radio Environment Map
RSU	RoadSide Unit
SAS	Spectrum Activity Surveillance
SI	Susceptible-Infected
SIC	Susceptible-Infected-Cured
SIR	Susceptible-Infected-Recovered
SIS	Susceptible-Infected-Susceptible
SU	Secondary User
VR	Virtual Reality
r.v.	random variable

Chapter 1
Introduction

1.1 Motivation

Data, which refers to *transmittable and storable computer information*, has been an integral part of modern society ever since the invention of computers. Especially in the past decade, its indispensable role in various applications, ranging from marketing [26] to scientific researches [7], has been re-enforced by advances in data mining, machine learning, and artificial intelligence (AI) studies. As the proliferation of smart devices that can interconnect with each other on-the-go, the creation, collection, and analysis of data are much easier and more accessible than before, leading to huge amounts of data being generated in both wired and wireless networks almost every second. For instance, the online social network (OSN) giant Facebook (now Meta) has 1.6 billion daily active users, who generate more than 4 PetaBytes new data every single day [27]. Meanwhile, the delivery networks of data are evolving into large and complex systems, due to the explosive growth of wireless devices, e.g., the number of Internet-of-Things (IoT) devices is expected to exceed 500 billion by 2030 [4], imposing a tangible impact on mobile data traffic. Consequently, recent years are witnessing the transition of data from a commodity owned by big companies, to a *service* that can be provided/acquired by practically anyone, just like the transition of computing resource in the cloud infrastructure a decade ago [22]. The principal course of such service is to *move* data from its generator(s) to its consumer(s) through a network of data carriers. In this data dissemination process, *data is mobile*, in the sense that both the traffic volume and whereabouts are constantly changing due to user movements and data forwarding actions.

J. Wang et al., *Encountering Mobile Data Dynamics in Heterogeneous Wireless Networks*, https://doi.org/10.1007/978-3-031-62906-8_1

1.1.1 Data Is Alive and Mobile

From the data owner or disseminator's perspective, it is their natural rights to know who have taken (temporary) possession of their data, where those data blocks have traveled to, and when a data block of interest stops circulating in a certain region. All of these questions are tied closely to movements of data during dissemination.

From the delivery network's perspective, data is *alive*, that is, interacting with individuals in the network, only when it is *mobile*. In other words, the *lifetime* of data begins at the time instant when it is first injected into the network, and stops when none of its copies is circulating in the network of interest any more. During its lifetime, every move of a data block induces dynamical changes in the network, with respect to both resources and network status. In the former aspect, mobile data utilizes various kinds of network resources, for example: storage resource is consumed at a device, when the data block is temporarily stored by intermediate data carriers (for later forwarding); bandwidth/spectrum resource is consumed, when it is transmitted between data carriers; computation resource is also consumed, when the data block needs to be fragmented, processed, or routed. On the other hand, operating status of both individuals and the networked system as a whole, e.g., capacity and system integrity, are in turn impacted by mobile data. For example, a computer malware, e.g., the SMS Trojan [8] that spreads over emails and messages, can hide in data blocks, piggyback on their dissemination processes, and attack multiple users in an institutional computer network. As such a process unfolds in time, normal operations of the networked system may no longer be sustained.

Therefore, it is both primitive and essential to understand data's *mobility*, for the design, management, and recovery of data-delivery networks, as well as the provision of service transparency to data owners. Specifically, we identify the following open questions to understand the dynamic processes with respect to mobile data:

1. *When* does a data block start and stop moving in a network?
2. *Where* is the data block of interest accessible during dissemination?
3. *How* do data blocks move in the heterogeneous wireless network?
4. *What* is the observable impact of mobile data on data-delivery networks?

Among these, the first and second questions focus on the dissemination process of a single data block, while the rest two questions are for cases of multiple blocks. Particularly, the first question focuses on the time domain, in which the cause, or the driving force of mobile data, dictates the start and stop of the dissemination process, i.e., the *lifetime* of a data block in a network. The second question focuses on the space domain, in which the *whereabouts* of data refer to the time-varying locations, where the data block (and its copies) is accessible. The third question asks for the *governing rule* of mobile data, taking interactions of multiple data dissemination processes into consideration. The last question focuses on the consequence and *impact* of data being mobile, especially on shared network resources.

Considering that movements of a single data block already create a dynamic process in the space that spans over time, geographical location, and spectrum domains, the sheer complexity of multiple data blocks being replicated, piggybacked and transmitted in the same network prohibits these questions to be answered with a single model, nor a simple solution. Therefore, we first specify and analyze the scenario of mobile data, and then introduce our solution that tackles the aforementioned problem in different domains.

1.1.2 Mobile Data Dynamics in Heterogeneous Wireless Networks

Thanks to developments in wireless communication and networking technologies, including 5G [24], dynamic spectrum access (DSA) [17], IoT [1], fog computing [25], and so on, wireless network has become the primary choice of many data disseminators, and the common practice of content delivery services as well. This phenomenon can be observed by the ever-increasing volume of wireless (especially mobile) traffic [5] and spectrum demand. For instance, ever since 2015, over 80% of social media traffic in the U.S. comes from wireless mobile devices, as well as over 50% of all website traffic worldwide [11]. Such a wireless data-delivery paradigm is adopted by various application scenarios, including data sharing/forwarding [13, 14] in Long Term Evolution (LTE) device-to-device communication (D2D) networks, mobile advertisement [21] in WiFi-LTE networks, safety message dissemination [14, 30] in vehicular networks, IoT provisioning by LTE-based fog [1], and many more. Motivated by its extensive applications, this book is devoted to *mobile data in heterogeneous wireless networks*, which are composed of both wireless end devices, such as mobile phones, smart vehicles, and sensory devices, and network elements of the wireless access network, such as base station (BS) in cellular networks, access point (AP) in wireless LAN, and roadside unit (RSU) in vehicular networks.

Such wireless data-delivery networks exhibit distinct characteristics, bringing new challenges to the field of networking: First, data carrying individuals (users) themselves are mobile, resulting in intermittent data transmission links and changing network topologies. Moreover, user mobility introduces the notion of 'where', i.e., geographical location, which further complicates the problem. Second, individuals in such networks can be highly diverse in many aspects, including communication protocol, radio access technology (RAT), data forwarding preference, mobility pattern, etc., creating a dynamic and heterogeneous environment, which is difficult to model and experiment on. Last but not least, the data-delivery network can be formed in a spontaneous manner, lacking possible control of any form, which further complicates the design, maintenance, and operation of such systems.

To address these challenges, we identify three dynamic processes, each of which describes the behavior of mobile data in one of the three domains, namely, *time*, *geographical space*, and *spectrum*, such that their properties are tractable to be analyzed individually, and collectively they are comprehensive enough to understand mobile data in heterogeneous wireless networks. These dynamic processes reflect the cause, manifestation, and result of data's mobility, respectively, and are hence referred to as *mobile data dynamics*.

1.1.2.1 Information Dynamics: The Driving Force of Mobile Data

Any data-delivery network is designed to facilitate the flow of *information*, in the form of moving data blocks, so the beginning and the end of information propagation decide the start and stop time of data movements. However, as networks evolve into more complex systems, increasing users of diverse backgrounds introduce information of different aspects, even *conflicting information*, e.g. rumor vs. truth, into delivery networks. Similar phenomena can also be observed in various forms of networked systems, if the information concept is generalized to include virus/malware, system operation status, and adoption of new products.

Despite different manifestations, we observe that all the conflicting information propagation instances share some common characteristics: the later-injected desired information targets at an existing undesired information, stops its propagation, and terminates a potential/ongoing *epidemic* of the latter. In this way, the undesired information resembles an infectious *virus* that can infect susceptible individuals, while the desired information functions as an *antidote* that can permanently immune susceptible individuals or cure infected individuals. Due to this asymmetry, the propagation process is *transient*, in the sense that its asymptotic behavior at time t goes to infinity is known (virus-free), and the two epidemic processes co-exist only for a short period of time, as opposed to the long-term coexistence (and equilibrium) in existing models on competing epidemics, e.g., [2, 6, 19].

Accordingly, a natural question is, how such propagation process evolves in *finite* time. In other words, there is little knowledge on the aftermath of conflicting information propagation via individual spreading, especially *when* the undesired information (virus) dies out, *how fast* the number of victims of the virus decreases below a predetermined level, and how to design effective information (antidote) distribution strategy to reduce the lifetime of undesired information in a network.

These questions have a broad impact on the design, operation, and management of networks, because information propagation is the driving force of mobile data, and the extinction of the undesired information marks the end of its propagation (generation of data traffic), while the impact of the propagation reveals requirements of data delivery services.

1.1.2.2 Coverage Dynamics: The Whereabouts of Mobile Data

In a wireless data-delivery network that includes mobile devices, a data dissemination process is actually a sequence of data relocation actions, which are the superposition of data transmissions and entity (data-carrying individuals) movements. In this scenario, data's mobility is manifested in its time-varying *coverage*, that is, the geographical location where the data block of interest can be accessed (legally by its consumers or monitors, and illegally by malicious attackers).

Then the direct question of mobile data follows: *where* is the data during a dissemination process? If such a dissemination process is viewed as a cause-and-effect phenomenon, existing researches mainly focus on single factors from the cause-direction, e.g., from the network topology and communication aspects. While most literature [14, 18, 21, 30] do not carry the notion of 'where', the ones that do [12, 36] assume entities are homogeneous in every aspect. To answer the *where* question for a heterogeneous scenario, a new model is needed to describe data's current whereabouts/coverage.

Knowing the dynamically changing data coverage is important to both the data owner(s) and the delivery network. To the former, data dissemination should be transparent to the customer (data owner/disseminator) as a service, while to the later, particularly network elements, including AP, BS, RSU, and gateways, changing coverage translates to traffic load and is hence relevant to resource management, as well as policy and charging plan design.

1.1.2.3 Spectrum Dynamics: Impact of Mobile Data in the Spectrum Domain

Due to the natural gap between its scarcity and high communication demand, radio *spectrum* is one of the most important resources in current wireless systems. Any data transmission in a wireless network will result in a spectrum *activity*, i.e., occupancy of a spectrum slice for a certain period of time (typically several milliseconds) at a geographical location, so that no other individual in this region can utilize the same slice simultaneously. In other words, *spectrum dynamics*, that is, time-varying spectrum activities over a geographical region, are the *outcome* of mobile data, as well as an impact to the system capacity.

To observe and evaluate such impact, it is necessary to have a spectrum surveillance system, which carries out continuous scans of spectrum activities on the frequencies of interest, for the purpose of usage data collection, including temporal and spatial patterns of spectrum occupancy, user mobility, as well as traffic patterns. Spectrum surveillance is particularly crucial to dynamic spectrum access (DSA)-enabled systems, because of the risks introduced by the open and opportunistic nature of DSA. In prior studies of spectrum surveillance strategies (e.g., [10, 15, 23]), an implicit assumption is that spectrum monitors are sufficiently powerful, such that they can watch over the entire geographical region of interest and tune/move without any limit. The fact is that most spectrum activities, including communications, attacks/jamming and monitoring/sniffing, are *local*, i.e., confined in both the spectrum domain and the space domain during a fixed-length time interval. This discrepancy is especially pronounced in wide-band wide-area spectrum monitoring, which naturally leads to an open question: *how to model* a spectrum surveillance process and *design* deployment strategies for monitors?

Answers to this question sketch a way to evaluate the impact of data mobility in emerging wireless systems, where DSA is enabled to improve spectrum efficiency.

1.1.2.4 Network-Data Interaction: Governing Rules of Mobile Data

In addition to the three mobile data dynamics, which are external manifestations of data's movements in different domains, it is equally necessary to study *how data move*, i.e., the governing rule of mobile data, to fully understand how their movements are constrained by resources available in a wireless system, and how their movements in turn affect the system operating status.

This question is especially meaningful to emerging networking paradigms, such as edge/fog computing [25], because they are expected to support frequent data offloading induced by IoT applications at a massive scale, with rather limited resources [16]. While existing studies in this line of research design offloading schemes to direct where data blocks should move, e.g., [3, 20, 32, 33, 35], to minimize the offloading latency [29] and/or energy consumption [28] for a single task, or to maximize revenue of the network [31], there is little knowledge on how to evaluate the performance of such systems, particularly how much resource the offloading process consumes, especially how it scales with the network size. In fact, scalability has been recognized as one of the major design concerns for the fog paradigm [9], and the crux of the resource consumption problem [34], which motivates us to model the data offloading processes and study the performance evaluation problem for the resource-constrained edge networks.

1.2 Contents of This Book

With the three mobile data dynamics identified in different domains, and network-data interaction specified for data's mobility, this book discusses mobile data in a *top-down* manner: we start from the driving force (information dynamics) in the time domain, then move to the description of whereabouts (coverage dynamics) in the space domain, further to the governing rules of data movements, and finally arrive at the impact on networks (spectrum dynamics) in the spectrum domain. Each step addresses one question (*when*, *where*, *how*, and *what*) identified at the beginning of this chapter. Next, we briefly summarize the research questions to be discussed in each of the following chapters, and outline the freshly gained knowledge toward the understanding of mobile data.

1.2.1 Information Dynamics: When Data Start and Stop Moving

Information propagation is the driving force of mobile data. With respect to information dynamics, Chap. 2 studies the propagation process of conflicting information, which is expected to answer *when data start and stop moving*. Considering the resemblance between a pair of conflicting information and a pair of virus and antidote, we model such a propagation process as two competing epidemics. To address the open question of how virus and antidote epidemics evolve in the same network, specifically, *how much time* it takes the undesired information (virus) to stop propagating, we derive bounds for such dynamics, in terms of network size and initial conditions. In this process, we reveal the influence of network topology on the lifetime of the undesired information. Further, as applications of the established model, we propose information injection strategies to reduce the lifetime of the undesired information, and design an algorithm to infer the time-varying number of information adopters in the network. Our contributions toward the *when* (Question #1) question are summarized as follows.

- **Modeling**: We propose a *Susceptible-Infected-Cured (SIC)* epidemic model to study the propagation process of conflicting information in a network, which captures the competing-while-spreading effect of the virus and antidote information.
- **Metric**: We define *extinction time* and *half-life time* of the virus, to describe the lifetime of the undesired information, as well as to quantify the effectiveness of information injection as a countermeasure against the undesired information.
- **Findings**: We derive bounds of the extinction time and half-life time, for SIC dynamics in networks with arbitrary topology, revealing how topological properties change the scaling of virus' lifetime over the size of the network.
- **Application**: We propose a divide-and-conquer guideline for desired information injection, to effectively reduce the extinction time of the undesired information, and further design an inference algorithm, to estimate the number of information adopters during the competition between the two pieces of information, to better understand the real-time impact of the undesired information in real networks.

1.2.2 Coverage Dynamics: Where Data Are during Dissemination

Chapter 3 focuses on the *whereabouts* of a data block (and its copies) during its dissemination process. To answer *where data are*, we formally define and quantify data coverage, that is, where the data block is accessible, such that data coverage can be represented by a time-varying signal on a *graph*, which captures physical movements of individual users/entities, and embeds geographical locations. Analyzing snapshots of the graph signal reveals the impact of user mobility, and how mobility can be used in predicting data coverage in the future. Our contributions toward the *whereabouts* (Question #2) of mobile data are summarized as follows.

- **Modeling**: We propose an entity model, based on which *data coverage* is formally defined, quantified by the *data-strength* metric, and formulated as a series of numeric signals on a graph, which embeds location information.
- **Metric**: We define the mobility dependence index (MDI) to quantitatively reveal the impact of user mobility on the change of data coverage.
- **Finding**: We find that the impact of user mobility is high, especially when entities are less active forwarding or the maximum moving speed is low, indicating possibility of predicting data coverage with long-term mobility traces.

1.2.3 Governing Rules: How Data Move during Offloading

Chapter 4 discusses the *governing rule* and the *impact* of mobile data in an offloading process, which is an exemplified application of the fog computing paradigm. With respect to *how data blocks move*, we study the inter-play of multiple tasks that compete for network resource during their offloading processes, and propose a gravity-based offloading model that can describe a variety of offloading criteria. With respect to the *impact on network resource*, we define device and network efforts, which quantify the amount of resource consumed during the offloading of any task. Our contributions toward the *governing rule* (Question #3) and *impact* (Question #4) of mobile data are summarized as follows.

- **Modeling**: We propose a *gravity* offloading model, to capture probabilistic task offloading processes, by which a variety of offloading criteria can be described.
- **Metric**: We define task *lifetime*, *device effort* and *network effort*, to quantify the offloading performance for individual tasks, and the amount of storage resource and communication resource consumed by the offloading process.
- **Findings**: We derive upper bounds for the performance metrics under the generic gravity rule, and find that the lifetime and device effort of individual tasks decrease as $O(\frac{1}{N})$ over the network size N, while the network effort decreases much slower, even remain $O(1)$ when load-balancing is adopted, indicating a possible resource drain in wireless edge networks.

1.2.4 Spectrum Dynamics: What Observable Impact Mobile Data Cause

Chapter 5 focuses on spectrum dynamics, which is the result of mobile data. Particularly, we discuss spectrum activity surveillance (SAS) processes over a large geographical region, to answer *what is the observable impact of mobile data* in the spectrum domain. To be more specific, we identify the goals of SAS to be sweep-coverage of the spectrum/space, and detection of spectrum culprits. Taking geographical space, and locality of spectrum activities into consideration, we propose a two-step solution, such that any SAS process can be formulated as a graph walk process. Two typical surveillance scenarios, namely, the dedicated and crowdsorucing scenarios, are analyzed to address their distinct design concerns, i.e., efficient sweep-coverage and quick detection of culprits, respectively. As an application of the proposed modeling approach, we propose deterministic and randomized monitor deployment strategies, whose performances, i.e., coverage time and detection time, are analyzed. Our contributions toward the *observable impact* (Question #4) of mobile data are summarized as follows.

- **Modeling**: We model spectrum activities in a *spectra-location space* that incorporates spectra, temporal and geographical domains, to capture their *locality*. Then the SAS strategy design is formulated as a graph walk problem.
- **Metrics**: We define the *coverage time* and *detection time* for monitor deployment strategies, so that the qualitative data collection and culprit detection objectives are translated to quantitative metrics, by which strategies can be fairly compared.
- **Findings**: We show by analysis and evaluation that, despite the switching capacity limits, randomized strategies are able to overcome the 'wandering hole' problem induced by adversarial culprits, and achieve guaranteed sweep-coverage and detection performances.

1.3 Organization of This Book

We hope this systematic study of mobile data dynamics expands our knowledge on data dissemination services as they migrate to the edge network, and provides practical guidelines for fast, efficient, and predictable service provisioning in heterogeneous wireless networks.

This book revolves around the cause, description, governing rule, and impact of mobile data, and takes a modeling approach to address the *when*, *where*, *how*, and *what* questions of mobile data, which are identified at the beginning of this chapter. The rest of this book is organized as follows: Chap. 2 discusses the conflicting information propagation process, to answer when data starts and stops moving in networks. Then in Chap. 3, coverage dynamics, that is the manifestation of mobile data in a wireless environment, is introduced to examine the whereabouts of a single data block (and its copies) in its dissemination process. Chapter 4 focuses on the governing rule of mobile data, which studies how multiple data blocks move during offloading, under resource constraints in a fog computing system. Chapter 5 focuses on observing the impact of mobile data, for which spectrum activity surveillance (SAS) is studied for DSA-enabled wireless systems. Finally, we discuss mobile data in the context of Edge Intelligence (EI) in Chap. 6, where we conclude the book and discuss open questions to inspire further research.

References

1. Ai, Y., Peng, M., Zhang, K.: Edge computing technologies for internet of things: a primer. Digital Commun. Netw. **4**(2), 77–86 (2018). https://doi.org/10.1016/j.dcan.2017.07.001. http://www.sciencedirect.com/science/article/pii/S2352864817301335
2. Beutel, A., Prakash, B.A., Rosenfeld, R., Faloutsos, C.: Interacting viruses in networks: Can both survive? In: Proceedings of the 18th ACM SIGKDD International Conference on Knowledge Discovery and Data Mining, KDD '12, pp. 426–434. ACM (2012). https://doi.org/10.1145/2339530.2339601. http://doi.acm.org/10.1145/2339530.2339601
3. Chen, L., Xu, J.: Socially trusted collaborative edge computing in ultra dense networks. In: Proceedings of the Second ACM/IEEE Symposium on Edge Computing, SEC '17, pp. 9:1–9:11. ACM, New York (2017). https://doi.org/10.1145/3132211.3134451. http://doi.acm.org/10.1145/3132211.3134451
4. Cisco Visual Networking Index: Global Mobile Data Traffic Forecast Update, 2017–2022. Technical Report (2019)
5. Cisco Visual Networking Index: Forecast and Trends, 2017–2022 White Paper. Technical Report, Cisco and/or its affliates (2019)
6. Dadlani, A., Kumar, M.S., Maddi, M.G., Kim, K.: Mean-field dynamics of inter-switching memes competing over multiplex social networks. IEEE Commun. Lett. **21**(5), 967–970 (2017). https://doi.org/10.1109/LCOMM.2017.2651815
7. Felt, M.: Social media and the social sciences: how researchers employ big data analytics. Big Data Soc. **3**(1), 2053951716645828 (2016). https://doi.org/10.1177/2053951716645828. http://dx.doi.org/10.1177/2053951716645828
8. Garnaeva, M., Chebyshev, V., Makrushin, D., Unuchek, R., Ivanov, A.: Kaspersky security bulletin 2014. overall statistics for 2014 (2014). http://securelist.com/analysis/kaspersky-security-bulletin/68010/kaspersky-security-bulletin-2014-overall-statistics-for-2014/
9. IEEE Std 1934-2018: Ieee standard for adoption of openfog reference architecture for fog computing, pp. 1–176 (2018). https://doi.org/10.1109/IEEESTD.2018.8423800
10. Jin, X., Sun, J., Zhang, R., Zhang, Y., Zhang, C.: Specguard: Spectrum misuse detection in dynamic spectrum access systems. In: 2015 IEEE Conference on Computer Communications (INFOCOM), pp. 172–180 (2015). https://doi.org/10.1109/INFOCOM.2015.7218380
11. Lella, A., Lipsman, A.: The 2016 u.s. cross-platform future in focus (comscore whitepaper) (2016). https://www.comscore.com/Insights/Presentations-and-Whitepapers/2016/2016-US-Cross-Platform-Future-in-Focus
12. Li, Y., Wang, W.: Horizon on the move: Geocast in intermittently connected vehicular ad hoc networks. In: 2013 Proceedings IEEE INFOCOM, pp. 2553–2561 (2013). https://doi.org/10.1109/INFCOM.2013.6567062
13. Li, Y., Wang, W.: Can mobile cloudlets support mobile applications? In: IEEE INFOCOM 2014 - IEEE Conference on Computer Communications, pp. 1060–1068 (2014). https://doi.org/10.1109/INFOCOM.2014.6848036
14. Li, Z., Wang, C., Yang, S., Jiang, C., Stojmenovic, I.: Space-crossing: Community-based data forwarding in mobile social networks under the hybrid communication architecture. IEEE Trans. Wirel. Commun. **14**(9), 4720–4727 (2015). https://doi.org/10.1109/TWC.2015.2424965
15. Li, M., Yang, D., Lin, J., Li, M., Tang, J.: Specwatch: Adversarial spectrum usage monitoring in crns with unknown statistics. In: IEEE INFOCOM 2016 - The 35th Annual IEEE International Conference on Computer Communications, pp. 1–9 (2016). https://doi.org/10.1109/INFOCOM.2016.7524555
16. Mao, Y., You, C., Zhang, J., Huang, K., Letaief, K.B.: A survey on mobile edge computing: The communication perspective. IEEE Commun. Surv. Tutor. **19**(4), 2322–2358 (2017). https://doi.org/10.1109/COMST.2017.2745201

17. Nika, A., Zhang, Z., Zhou, X., Zhao, B.Y., Zheng, H.: Towards commoditized real-time spectrum monitoring. In: Proceedings of the 1st ACM Workshop on Hot Topics in Wireless, HotWireless '14, pp. 25–30. ACM, New York (2014). https://doi.org/10.1145/2643614.2643615. http://doi.acm.org/10.1145/2643614.2643615
18. Pastor-Satorras, R., Castellano, C., Van Mieghem, P., Vespignani, A.: Epidemic processes in complex networks. Rev. Mod. Phys. **87**, 925–979 (2015). https://doi.org/10.1103/RevModPhys.87.925
19. Prakash, B.A., Beutel, A., Rosenfeld, R., Faloutsos, C.: Winner takes all: Competing viruses or ideas on fair-play networks. In: Proceedings of the 21st International Conference on World Wide Web, WWW '12, pp. 1037–1046. ACM, New York (2012). https://doi.org/10.1145/2187836.2187975. http://doi.acm.org/10.1145/2187836.2187975
20. Pu, L., Chen, X., Xu, J., Fu, X.: D2d fogging: An energy-efficient and incentive-aware task offloading framework via network-assisted d2d collaboration. IEEE J. Sel. Areas Commun. **34**(12), 3887–3901 (2016). https://doi.org/10.1109/JSAC.2016.2624118
21. Qin, J., Zhu, H., Zhu, Y., Lu, L., Xue, G., Li, M.: Post: Exploiting dynamic sociality for mobile advertising in vehicular networks. IEEE Trans. Parallel Distr. Syst. **27**(6), 1770–1782 (2016). https://doi.org/10.1109/TPDS.2015.2467392
22. Ren, X., London, P., Ziani, J., Wierman, A.: Datum: Managing data purchasing and data placement in a geo-distributed data market. IEEE/ACM Trans. Netw. **26**(2), 893–905 (2018). https://doi.org/10.1109/TNET.2018.2811374
23. Shin, D.H., Bagchi, S.: Optimal monitoring in multi-channel multi-radio wireless mesh networks. In: Proceedings of the Tenth ACM International Symposium on Mobile Ad Hoc Networking and Computing, MobiHoc '09, pp. 229–238. ACM, New York (2009). https://doi.org/10.1145/1530748.1530780. http://doi.acm.org/10.1145/1530748.1530780
24. Tehrani, M.N., Uysal, M., Yanikomeroglu, H.: Device-to-device communication in 5g cellular networks: challenges, solutions, and future directions. IEEE Commun. Mag. **52**(5), 86–92 (2014). https://doi.org/10.1109/MCOM.2014.6815897
25. Vaquero, L.M., Rodero-Merino, L.: Finding your way in the fog: Towards a comprehensive definition of fog computing. SIGCOMM Comput. Commun. Rev. **44**(5), 27–32 (2014). https://doi.org/10.1145/2677046.2677052. http://doi.acm.org/10.1145/2677046.2677052
26. Wedel, M., Kannan, P.: Marketing analytics for data-rich environments. J. Market. **80**(6), 97–121 (2016). https://doi.org/10.1509/jm.15.0413
27. Wiener, J., Bronson, N.: Facebook's top open data problems (2014). https://research.fb.com/facebook-s-top-open-data-problems/
28. Wu, H.: Performance modeling of delayed offloading in mobile wireless environments with failures. IEEE Commun. Lett. **22**(11), 2334–2337 (2018). https://doi.org/10.1109/LCOMM.2018.2866585
29. Xiao, Y., Krunz, M.: Qoe and power efficiency tradeoff for fog computing networks with fog node cooperation. In: IEEE INFOCOM 2017 - IEEE Conference on Computer Communications, pp. 1–9 (2017). https://doi.org/10.1109/INFOCOM.2017.8057196
30. Yan, T., Zhang, W., Wang, G.: Dove: Data dissemination to a desired number of receivers in vanet. IEEE Trans. Veh. Technol. **63**(4), 1903–1916 (2014). https://doi.org/10.1109/TVT.2013.2287692
31. Yang, Y., Chang, X., Han, Z., Li, L.: Delay-aware secure computation offloading mechanism in a fog-cloud framework. In: 2018 IEEE Intl Conference on Parallel Distributed Processing with Applications, Ubiquitous Computing Communications, Big Data Cloud Computing, Social Computing Networking, Sustainable Computing Communications (ISPA/IUCC/BDCloud/SocialCom/SustainCom), pp. 346–353 (2018). https://doi.org/10.1109/BDCloud.2018.00061
32. Yang, Y., Wang, K., Zhang, G., Chen, X., Luo, X., Zhou, M.: Meets: Maximal energy efficient task scheduling in homogeneous fog networks. IEEE Internet Things J. **5**(5), 4076–4087 (2018). https://doi.org/10.1109/JIOT.2018.2846644

33. Yao, J., Ansari, N.: Reliability-aware fog resource provisioning for deadline-driven iot services. In: 2018 IEEE Global Communications Conference (GLOBECOM), pp. 1–6 (2018). https://doi.org/10.1109/GLOCOM.2018.8647378
34. Yousefpour, A., Fung, C., Nguyen, T., Kadiyala, K., Jalali, F., Niakanlahiji, A., Kong, J., Jue, J.P.: All one needs to know about fog computing and related edge computing paradigms: a complete survey. J. Syst. Archit. (2019). https://doi.org/10.1016/j.sysarc.2019.02.009. http://www.sciencedirect.com/science/article/pii/S1383762118306349
35. Zhang, H., Xiao, Y., Bu, S., Niyato, D., Yu, F.R., Han, Z.: Computing resource allocation in three-tier iot fog networks: a joint optimization approach combining stackelberg game and matching. IEEE Internet Things J. **4**(5), 1204–1215 (2017). https://doi.org/10.1109/JIOT.2017.2688925
36. Zhang, H., Zhang, Z., Dai, H.: Gossip-based information spreading in mobile networks. IEEE Trans. Wirel. Commun. **12**(11), 5918–5928 (2013). https://doi.org/10.1109/TWC.2013.100113.130619

Chapter 2
Information Dynamics: Modeling and Analysis of Conflicting Information Propagation in a Finite Time Horizon

Information propagation is the driving force of mobile data dynamics. In this chapter, we study information dynamics, particularly the propagation process of conflicting information in networks, to provide in-depth understanding of how network topology determines the lifetime of mobile data dynamics. Through modeling, analysis, and simulation, we find that the lifetime of the undesired information can be upper and lower bounded by functions of the network size and topological properties. Specifically, when connections concentrate on a few individuals to create bottlenecks in networks, the lifetime of the undesired information can change from decreasing to increasing with the network size. Taking computation complexity into consideration, we obtain upper bounds with vertex eccentricities, that is, the largest distance between an individual and any other individuals in the network, from which we propose a 'divide-and-conquer' guideline to inject desired information, such that the undesired information can be eliminated in a short period of time. In addition, to observe the propagation process in a finer time resolution, we propose an inference algorithm to estimate the number of information adopters, which can foresee the instantaneous evolution of the dynamics before it fully unfolds, providing valuable information for the prevention and control of undesired information propagation.

2.1 Motivation and Related Work

Due to proliferating mobile devices and emerging mobile applications, people are more connected with each other than ever. Consequently, networked systems, such as OSN, institutional computer networks, and IoT, are developing into much larger and more complex structures than before. For instance, the OSN giant Facebook has 1.6 billion daily active users, who generate more than 4 PetaBytes new data every day [40], while the number of IoT devices is expected to exceed 500 billion by 2030 [12], imposing a tangible impact on mobile data traffic. As more and

J. Wang et al., *Encountering Mobile Data Dynamics in Heterogeneous Wireless Networks*, https://doi.org/10.1007/978-3-031-62906-8_2

more individuals, e.g., users in OSN, and devices in IoT, join such systems, inconsistent, even *conflicting information* is injected into the network during the same time period, leading to an interesting competition while both pieces propagate via individual connections.

By *conflicting*, we mean two pieces of information that can not be admitted by the same individual at the same time. For example, it is highly unlikely, for an OSN user to simultaneously admit the truth and a rumor that contradicts with the truth. Particularly for such pairs, in which one piece (the desired information, e.g., truth) is apparently more credible than the other (the undesired information, e.g., a rumor), an individual who has chosen to admit the desired information, will not be affected by its undesired counterpart. Therefore, the undesired information, though spreading via individual connections itself, will be eliminated from the network given sufficient time. This phenomenon can be seen in various systems.

2.1.1 *Motivating Examples in Different Systems*

Conflicting information propagation is prevalent in social networks, e.g., the word-of-mouth networks among acquaintances, and OSN, which has become the arena of clashing opinions, unverified reports, and publicity campaigns. Similar competing-while-spreading phenomena can also be observed in engineered systems, such as the computer network of an institution, and IoT.

2.1.1.1 Rumor vs. Truth in OSN

After the Boston bombing incident on April 15th, 2013, Reddit users started an online suspect hunt, which identified an innocent person as the bomber [1]. This rumor (undesired information in the form of image data) spread rapidly on both Reddit and Twitter, leading to serious cyber-harassment to the wrongly-accused. The number of mentions regarding this rumor quickly decreased after the police released correct information (desired information) [22]. Similarly in the OSN, the meme tracker App [41] recorded the number of posts with "#SpecialOlympics#" from March 20th to 24th, 2009, as shown in Fig. 2.1. At that time, U.S. President Obama made an inappropriate joke in 'The Tonight Show with Jay Leno', and soon apologized to the Special Olympics chairman to correct his mistake. However, it took four days for this incident to die down completely, after the apology was made. In case of such epidemic spreading of both desired and undesired information, a natural question is: *when will the undesired information stop propagating*, such that it will not affect users in the network any more?

Moreover, it is observed that the lifetimes (often referred to as the news cycles) of both rumors and news are becoming noticeably shorter [24, 41] than before. Is this phenomenon caused by the rapid increase of active users in the OSN [40]? What

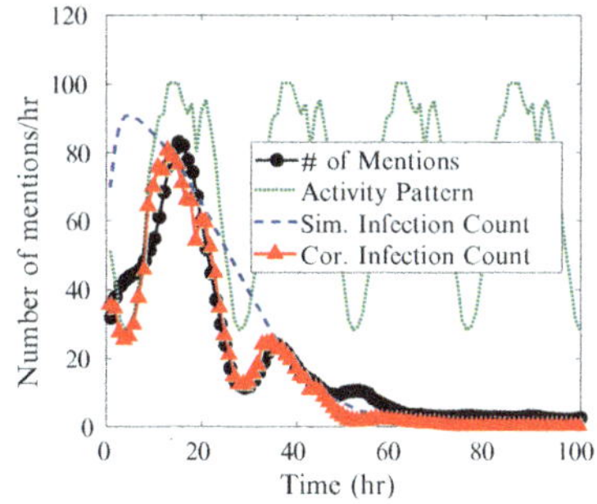

Fig. 2.1 Motivating example #1: Number of mentions for the '*Special Olympics*' incident in an OSN

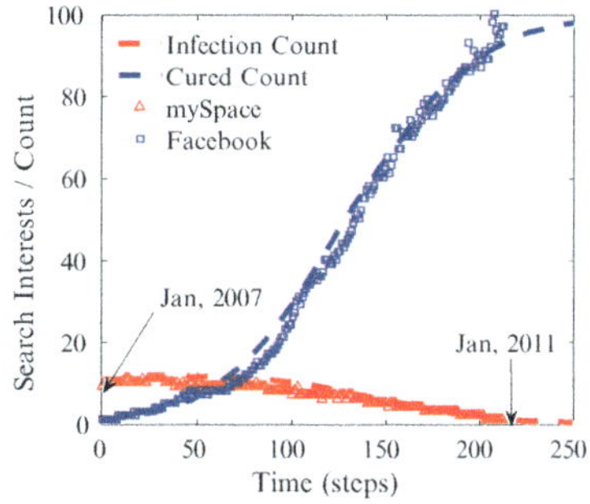

Fig. 2.2 Motivating example #2: Popularity of two rivalry OSN products, MySpace vs. Facebook(Meta)

does this phenomenon imply about the topological structure of the OSN, which is too big and complex to be studied as a whole?

2.1.1.2 Advanced Product vs. Outdated Product in Word-of-Mouth Networks

When two rivalry products compete through the word-of-mouth network, where people are influenced by the 'reputation' of a product among their friends, the newer and more advanced product will eventually dominate the market by gradually swapping out its outdated counterpart that is less appealing to customers. For instance, Facebook (Meta) overturned the OSN market in its competition with MySpace, which was quite popular when Facebook just opened registration in January 2007, as shown in Fig. 2.2. The newer mobile game Clash of Clans, drew mobile users' attention quickly from its counterpart, Candy Crush, as shown in Fig. 2.3. In case of rivalry products' competition, the manufacturer of the better product would be interested in the market share (adopter of the desired information) its product takes before reaching a full market penetration, to plan and allocate business resources. They are therefore interested in the following question, that is, how does the dynamic process evolve, particularly *how the number of adopters of different information changes* over time?

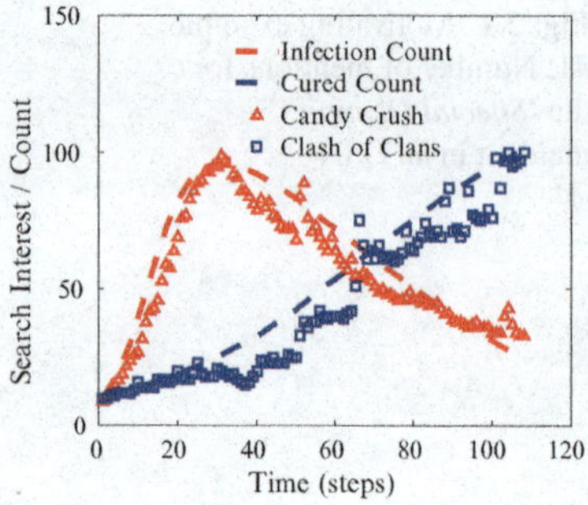

Fig. 2.3 Motivating example #3: Popularity of two mobile games, Candy Crash vs. Clash of Clans

2.1.1.3 Malware vs. Security Patch in Institutional Computer Networks and Faults vs. Restoration Commands in IoT

Unlike social networks, large engineered systems such as institutional computer networks [38] and IoT usually operate under the control of system administrators. When the system administrator of an institutional computer network spots spreading malware infections, e.g., SMS Trojans [17] that spread over emails and messages, and the Chameleon virus [29] that spreads over WiFi links, a possible countermeasure is to distribute replicable security patches, which can fix infected computers, and also prevent uninfected computers from malware infections. Similarly in IoT systems, such as a smart grid, when cascading failures [39] are triggered by overloaded stations, the administrator can apply load shedding measures [4] to restore some affected stations, such that their neighboring stations can also benefit from the restoration, and gradually return to normal operation. In both cases, the system administrator knows the structure of the networked system, and hence may proactively inject desired information (security patch, load shedding command), to control and eliminate the epidemic spread of undesired information (malware, fault). Consequently, a practical question for the system administrator is, *where to inject the desired information*, such that the undesired information (malware, fault or failure) can be eliminated faster? In other words, how to leverage knowledge on network topology to design more effective countermeasures against the viral spreading of undesired information?

Despite the manifestation, all these conflicting information propagation processes have two defining characteristics: (1) There are *two* pieces of information circulating the same network, both spread via contacts of individuals in an *epidemic* manner. (2) The later-injected desired information can convert victims of the undesired information back (to normal states), just like a replicable *antidote* can cure or immune an individual from an infectious *virus*, as a result of which, the cured individual will not be infected by the same virus again, but not vice versa.

2.1.2 Related Work

Considering the great resemblance between information propagation and virus spreading, *epidemic models* have been adopted to describe the propagation process over individual connections, e.g., rumors spread between friends, and malwares spread between computers, in which the information is modeled as an infectious virus. Literature on this line of research can be categorized into single-virus epidemics and multi-virus (or competing) epidemics.

Among the extensive study on single-virus epidemics [8, 16, 21, 23, 32, 33, 39], Ganesh et al. [16] and Krishnasamy et al. [21] identified the significant impact of the Cheeger constant $\eta(\mathcal{G})$, which measures the expansion property of the underlying network $\mathcal{G}$, on the propagation time of the virus. To be more specific, the spreading time of the virus is upper bounded by a function of $\eta(\mathcal{G})$ [21] under the Susceptible-Infected (SI) model, while for the Susceptible-Infected-Susceptible (SIS) model that allows infected individuals to recover on themselves, the virus can live for an exponentially long time with respect to the size n of the network [16], if the recovery-to-infection rate is larger than $\eta(\mathcal{G})$. However, these single-virus models can not describe the competition between multiple pieces of information, which is the key characteristic of conflicting information propagation. In fact, the single-virus SI epidemics can be viewed as a special case of the conflicting information propagation process, in the sense that it only observes the epidemic process of the antidote, which is not affected by the virus.

With respect to competing epidemics between conflicting information, existing literature can be further categorized into *population dynamics* and *network dynamics* [31], depending on whether network topology is taking into consideration. In population dynamics, participants of the propagation process are assumed to be a well-mixed population, i.e., there is no notion of network, which does not apply to most propagation scenarios. In contrast, network dynamics view participants of the propagation process as a heterogeneous group, and model their connections with a graph structure. Among these, Lin et al. [26] utilized mean field approximation (MFA) to conduct asymptomatic and numerical analysis of the propagation process. Prakash et al. [34] proposed an SI_1I_2S model, and proved that a piece of information with faster propagation speed will eventually eliminate its slower counterpart as time approaches infinity. More recently, Dadlani et al. modeled competing memes as epidemic processes on multi-layered graphs, and derived critical survival threshold of a meme [13] to be persistent. Newman [30] found the coexistence threshold of two competing epidemics on networks with known degree distributions, under the Susceptible-Infected-Recovered (SIR) model. From the perspective of propagation model, both the linear threshold model and SI_1I_2S model allow any individual to switch back and forth between different pieces of information, in which context the asymptotic behavior (steady state of the system as time $t \to \infty$) of the dynamics is of more interest. In our case, however, the desired information is much more credible than its undesired counterpart, so competition between the two only takes place in *finite time*, for which existing analysis on asymptotic behavior does not apply.

2.1.3 Our Approach and Contributions

Motivated by the lack of study, this chapter discusses the propagation process of a pair of virus-antidote like conflicting information in networked systems, in order to understand *when* the resulted mobile data dynamics starts and stops in such systems. Our contributions can be summarized as follows.

We propose a *Susceptible-Infected-Cured (SIC)* propagation model, to capture the competing and epidemic spreading nature of a conflicting information pair, and identify two pivots in the lifetime of the undesired information, namely the *extinction time* τ_e and *half-life time* $\tau_{\frac{1}{2}}$, to quantify the dynamic evolution in the time domain.

We find both lifetime metrics are upper bounded by functions of $\eta(\mathcal{G})$, the Cheeger constant measuring the level of bottleneck-ness of $\mathcal{G}$, which indicates the lifetime of the undesired information does not always decrease with the network size n. When edges of the network become less and more concentrated such that $\eta(\mathcal{G}) = O(\log n)$, the lifetime of the virus will decrease with n, as exemplified by the dichotomy of $\mathbb{E}(\tau_e)$ and $\mathbb{E}(\tau_{\frac{1}{2}})$ for two extreme topologies, i.e., clique and star.

Considering that $\eta(\mathcal{G})$ is difficult to obtain (NP-complete) for general networks, we show that the lifetime of the undesired information can also be upper and lower bounded by functions of vertex eccentricities, which can be obtained with fast algorithms [37]. These bounds not only enable us to estimate the lifetime metrics, but also imply a practical 'divide-and-conquer' guideline for information injection to be used as a countermeasure against viral spreading.

We design an inference algorithm to estimate the number of information adopters at any time, given the initial condition and the topology of the network, which can be used to observe and predict the dynamic evolution with a finer resolution in time.

2.2 Preliminaries and Problem Statement

In this section, we introduce terminologies, assumptions and definitions of the Susceptible-Infected-Cured propagation model, which captures the competition between a pair of *conflicting* information, referred to as a *virus* and an *antidote*. Then we formally define *lifetime* metrics of the virus, so as to formulate the *when* problem. First, we specify the scope of conflicting information studied in this chapter.

2.2.1 Conflicting Information Pair: Virus *x* and Antidote a_x

Conflicting information has been defined as "two pieces of texts that are extremely unlikely to be considered true simultaneously" in relevant information collection studies [15]. Considering that information takes various forms other than text, such as source code, operating status of devices, and commands in the motivating examples, we extend the definition of conflicting information to *mutually-exclusive* information that can not be possessed/admitted by the same individual at the same time. Particularly, we focus on a *virus-antidote* pair, in which the desired information (referred to as antidote a_x) is of dominant credibility/power over its undesired counterpart (referred to as virus x). In other words, the antidote a_x will kill virus x, if they are both present on the same vertex, but not vice versa. Consequently, virus x can not re-infect a cured individual, who has admitted antidote a_x. This is different from the symmetric setting in other models [3, 34], where antidote a_x is treated as another virus that is symmetric to x, instead of an antidote to x.

The rationale behind the asymmetry in our model comes from observations in real-world examples and modeling accuracy concerns. First, for the case of security patch vs. computer malware, and restoration commands vs. (cascading) faulty status, the desired information (e.g., security patch) is injected purposely by the system to eliminate the undesired information (e.g., malware), and it is only reasonable that a malware can not leverage a fixed bug to attack the system. Second, for the case of clashing opinions and adoption of different products, once an individual is convinced(infected) by the newer and better product a_x, it is unlikely for him/her to switch back to the older product x, unless the older product has an upgrade (to the newest version) x'. These new injections of x' are considered as the beginning of a new epidemic process with x' as the antidote to virus a_x in our model. In the SI_1I_2S model [3, 34], x' and x are considered as the same virus competing with a_x, such that re-infection (or switching-back from a_x to x) is allowed, because of its focus on long-term (time $t \rightarrow \infty$) behaviors. A drawback of the existing approach is that x' and x are automatically associated with the same propagation speed. On the contrary, our model allows x, a_x, x' to have different propagation speeds, capturing competitions of both a_x vs. x, and x' vs. a_x, and is hence more accurate in modeling.

2.2.2 Network $\mathcal{G}(\mathcal{V}, \mathcal{E})$

The underlying *network* of information (virus x and antidote a_x) propagation is described as a graph $\mathcal{G}(\mathcal{V}, \mathcal{E})$, where vertex set $\mathcal{V}$ corresponds to the set of individuals, such as devices in IoT, while edge set $\mathcal{E}$ corresponds to the set of connections between any of two individuals, such as the communication links between two devices. An edge $e(i, j)$ exists, if vertex i and j can directly exchange information. For any vertex $v \in \mathcal{V}$, its *neighborhood* $\mathcal{N}(v) := \{u \in \mathcal{V} | (u, v) \in \mathcal{E}\}$ is defined as the set of vertices that are directly connected to vertex v. The topology of the network can be described by its adjacency matrix $\mathbf{A}_{n \times n} = (a_{u,v})_{u,v \in \mathcal{V}}$, where $a_{u,v} = a_{v,u} = 1$, if there exits an edge $e(u, v) \in \mathcal{E}$ between the two vertices u and v. We make the following assumptions of network $\mathcal{G}$:

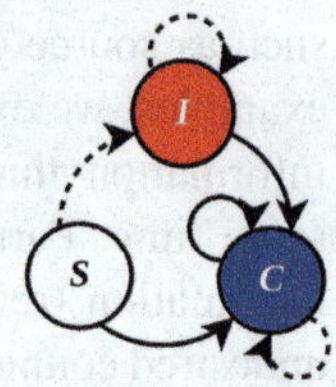

Fig. 2.4 State transitions of individual vertices in the SIC propagation model

1. *Undirected*: Edge $e(i, j) = e(j, i)$ identifies mutual and bidirectional connections between vertices i and j.
2. *Connected*: Information (x and a_x) can spread to every vertex in $\mathcal{V}$.
3. *Static*: Both size $n := |\mathcal{V}|$ and the adjacency matrix $\mathbf{A}$ of the network remain the same during the epidemic evolution.[1]

2.2.3 Epidemic Propagation Process

Both virus x and antidote a_x spread in an *epidemic* manner on network $\mathcal{G}$. To describe this epidemic process, each vertex is associated with a *state* that changes over time.

2.2.3.1 State Transitions

Let r.v. $X_v^x(t) : \Omega \to \Lambda = \{0, 1, -1\}$ denote the state of vertex $v \in \mathcal{V}$ at time t. Values of X_v^x correspond to different nodal states, as shown in Fig. 2.4.

Susceptible at Time t The default state $X_v^x(t) = 0$ (white circle with letter S in Fig. 2.4) indicates that neither the virus x nor the antidote a_x has reached vertex v by time t, so it is possible for v to be infected by x, or cured/immunized by a_x in the future, if any of them propagates to vertex v via contacts with other vertices.

Infected at Time t_i If a copy of virus x reaches susceptible vertex v at t_i, v becomes infected at t_i, which implies $X_v^x(t_i) = 1$ and $\lim_{t \to t_i^-} X_v^x(t) = 0$. This *infect* action is shown by the dashed arrow from susceptible state to the *infected* state (red circle with letter I) in Fig. 2.4. At this state, vertex v will try to *infect*, i.e., pass copies of virus x to any u of its susceptible neighbors, $\mathcal{N}_S^x(t_i, v) = \{u \in \mathcal{N}(v) | X_u^x(t_i) = 0\}$, after a random period of time $s_v^x(u)$. Vertex v will stay in infected state until it receives a copy of antidote a_x.

[1] Though connection links in heterogeneous wireless network scenarios may be unstable, the social connection that directs information exchange is less likely to change during the information propagation process. Therefore, throughout this chapter, we assume the network $\mathcal{G}$ to be static.

Cured at Time t_c If at t_c, a copy of antidote a_x reaches vertex v for the first time, i.e., $\lim_{t \to t_c^-} X_v^x(t) \geq 0$, the state of v changes to *cured* at t_c, that is, $X_v^x(t) = -1$, as shown as the blue circle with letter C in Fig. 2.4. At this state, vertex v will pass copies of antidote to any u of its neighbors $\mathcal{N}_{NC}^x(t, v) = \{u \in \mathcal{N}(v) | X_u^x(t_i) \geq 0\}$ after a random period of time $s_v^{a_x}(u)$. Vertex v will stay cured for the rest of the time, i.e., $X_v^x(t) = -1$ for any $t > t_c$.

2.2.3.2 Propagation Rules

As shown in Fig. 2.4, the state transitions of vertex v are driven by an infection event of virus x, or a curing event of antidote a_x, whose speeds are controlled by the infection intervals $\{s_u^x(v)\}_{v \in \mathcal{N}_S(t,u)}$, and the curing intervals $\{s_u^{a_x}(v)\}_{v \in \mathcal{N}_{NC}(t,u)}$, respectively. These intervals can be of arbitrary lengths, but to make this problem tractable, we follow the convention in [21, 34], and assume time homogeneity for the propagation process: For any vertex u, random intervals $\{s_u^x(v)\}_{v \in \mathcal{N}_S(t,u)}$ and $\{s_u^{a_x}(v)\}_{v \in \mathcal{N}_{NC}(t,u)}$ are two groups of r.v.'s satisfying (1) pairwise independent, and (2) exponentially distributed with parameters $\beta_{u,v}^x$ and $\gamma_{u,v}^x$, respectively.

From the perspective of time, $\beta_{u,v}^x = \beta_{v,u}^x$ is known as the *virulence* (or *infection rate*) of virus x, while $\gamma_{u,v}^x = \gamma_{v,u}^x$ as the *curing rate* of antidote a_x, representing how frequently a copy of virus x or antidote a_x is exchanged via edge $e(u, v)$, respectively. Their formal definitions are given as

$$\beta_{u,v}^x := \lim_{t \to 0^+} \frac{\mathbb{P}(s_u^x(v) \leq t)}{t}, \tag{2.1}$$

$$\gamma_{u,v}^x := \lim_{t \to 0^+} \frac{\mathbb{P}(s_u^{a_x}(v) \leq t)}{t}, \tag{2.2}$$

where $s_u^x(v)$ ($s_u^{a_x}(v)$, respectively) is the time period between the infection (curing) of u, and the time when u passes a copy of virus x (antidote a_x) to v via edge $e(u, v)$.

From the perspective of probability, $\beta_{u,v}^x$ is also known as the *infection probability* over unit time on edge $e(u, v)$, which can be explained by considering a simple network composed of two connected vertices, i.e., $\mathcal{V} = \{u, v\}$. In this case, at time t, given $X_v^x(t) = 0$, the probability that v gets infected by u in Δt is

$$\mathbb{P}\Big(X_v^x(t + \Delta t) = 1 | X_v^x(t) = 0\Big) = \Delta t \cdot \beta_{u,v}^x \cdot \mathbb{1}_{\{X_u^x(t)=1\}} + o(\Delta t), \tag{2.3}$$

where r.v. $s_u^x(v) \sim \text{Exp}(\beta_{u,v}^x)$, with mean $\mathbb{E}(s_u^x(v)) = \frac{1}{\beta_{u,v}^x}$. Similar results also apply to r.v. $s_u^{a_x}(v)$ and curing rate $\gamma_{u,v}^x$. Therefore, when the time interval is of unit length, i.e., $\Delta t = 1$, the state transition probability $\mathbb{P}\Big(X_v^x(t + \Delta t) = 1 | X_v^x(t) = 0\Big) = \mathbb{P}(s_u^x(v) \leq t) = \beta_{u,v}^x$ equals to the infection rate $\beta_{u,v}^x$ in value. This equation bridges the gap between continuous-time modeling/analysis and discrete-time simulations.

2.2.4 Problem Formulation

The SIC model in the previous subsection defines the propagation at individual vertex level. At the system level, evolution of the SIC dynamics can be described by the set of vertices that are susceptible, infected, and cured, respectively. This is because at any time t, vertex set $\mathcal{V} = \mathcal{S}^x(t) \cup \mathcal{I}^x(t) \cup \mathcal{C}^x(t)$, where the susceptible set $\mathcal{S}^x(t) := \{v \in \mathcal{V} | X_v^x(t) = 0\}$, the infected set $\mathcal{I}^x(t) = \{v \in \mathcal{V} | X_v^x(t) = 1\}$, and the cured set $\mathcal{C}^x(t) = \{v \in \mathcal{V} | X_v^x(t) = -1\}$ are mutually disjoint and form a partition of $\mathcal{V}$. Evolution of an SIC epidemic process can then be captured by the time-varying *infection count*, $I^x(t) := |\mathcal{I}^x(t)|$, and the *cured count*, $C^x(t) := |\mathcal{C}^x(t)|$. For the ease of notation, we suppress x, and write $X_v(t),\ \mathcal{S}(t),\ \mathcal{I}(t),\ \mathcal{C}(t)$ instead.

Figure 2.5 illustrates the evolution process of an SIC dynamics over a network with 12 vertices. As shown in Fig. 2.5a, before the antidote a_x is injected, $I(t_0^-) = 6$ vertices in $\mathcal{I}(t_0^-) = \{v_1, v_2, v_3, v_5, v_9, v_{11}\}$ are infected (colored in red). Then at t_0, one unit of antidote is given to vertex v_8, and cures it immediately (indicated as blue), that is, $C(t_0) = \{v_8\}$, as shown in Fig. 2.5b. As time t proceeds, states of the 12 vertices change as illustrated in Fig. 2.5c–e. Eventually, the virus dies out at time $t = t_0 + \tau_e$, and the infection count $I(t) = 0$.

As shown in the previous example (Fig. 2.5), the evolution of the SIC dynamics is recorded by the infection and cured counts over time. Based on these two system-level states, we define lifetime metrics of the virus to answer *when* the undesired information dies out. Specifically, we identify two pivots in the time domain, namely, the *extinction time* and *half-life time*, to quantify *how fast* the virus dies, which also indicates the effectiveness of an instance of antidote distribution.

Definition 2.1 For an SIC dynamic of virus x and antidote a_x, the **extinction time** of virus x, denoted as τ_e, is defined as the time interval between t_0 and the time instant that infected set $\mathcal{I}(t)$ becomes empty for the first time, that is,

$$\tau_e := \inf\{t > t_0 : \mathcal{I}(t) = \phi\} - t_0, \tag{2.4}$$

where t_0 is the time instant when C_0 copies of antidote are injected into network $\mathcal{G}$.

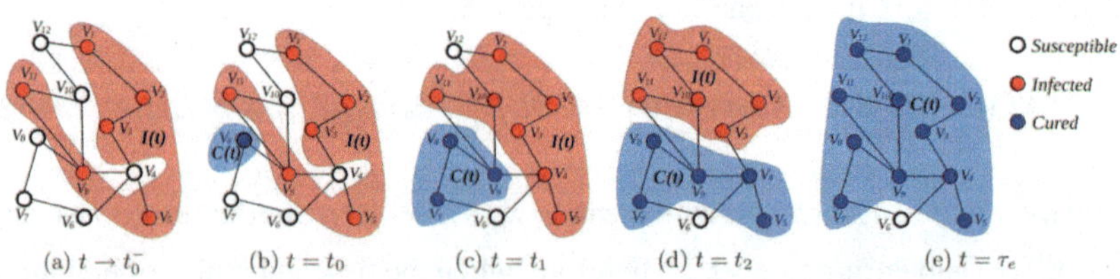

Fig. 2.5 An example of an SIC epidemics on a network of 12 vertices

The extinction time τ_e is a finite[2] r.v. on measurable space $(\Omega^n, 2^{\Omega^n}, \mathbb{P})$. It answers *when the virus dies out*, because time instant $t_0 + \tau_e$ marks the *end point* of the virus's life in $\mathcal{G}$. In other words, the infection count decreases from $I(t_0)$ to 0 during a time interval of length τ_e, so we know the virus (undesired information) dies out at an average speed of $\frac{I(t_0)}{\tau_e}$. But it is still not clear how such speed changes during the lifetime of the virus, i.e., *how fast* the virus dies out. This question is relevant to a lot of realistic questions on the impact of the virus, e.g., when will the majority of individuals be free of the undesired information? To answer this question, we identify another pivot in time, and define the half-life time of the virus.

Definition 2.2 For an SIC dynamic of virus x and antidote a_x, the **half-life time** of the virus epidemic, denoted as $\tau_{\frac{1}{2}}$, is defined as the time interval between t_0 and the last time that event $\{I(t) \geq \frac{1}{2}I(t_0)\}$ happens after t_0, that is,

$$\tau_{\frac{1}{2}} := \sup\{t \in [t_0, t_0 + \tau_e] : I(t) \geq \frac{1}{2}I(t_0)\} - t_0, \tag{2.5}$$

where $I(t_0) > 0$ is the initial infection count at t_0.

The half-life time $\tau_{\frac{1}{2}} : \Omega^n \rightarrow [t_0, t_0 + \tau_e]$ is also a finite r.v. on the same measurable space as r.v. τ_e. The term *half-life* is originally from Chemical Kinetics, which describes the decay of discrete entities. Unlike in Chemical Kinetics, where half-life is the mean, we define half-life as the actual time interval until event $\{I(t) \geq \frac{1}{2}I(t_0)\}$ happens for the *last* time. The physical meaning of $\tau_{\frac{1}{2}}$ can be understood as follows: the competition between the undesired information and its desired counterpart mainly takes place before pivot time $t_0 + \tau_{\frac{1}{2}}$, after which the undesired information can be deemed as controlled, because the number of its victims will never exceed the threshold $I(t_0)/2$. In this sense, the two lifetime metrics, extinction time τ_e and half-life time $\tau_{\frac{1}{2}}$, illustrate *how fast* the virus dies, because for a fixed initial condition, that is, $I(t_0)$ and $C(t_0)$, the larger the gap $\tau_e - \tau_{\frac{1}{2}}$, the faster the undesired information dies during its most hazardous phase $[t_0, t_0 + \tau_{\frac{1}{2}}]$.

To understand these two concepts, we present all the events according to the order of their occurrences in time, as illustrated in Fig. 2.6. In this diagram, a red arrow corresponds to an infection, and a blue one represents a curing event. At $t_0 + \tau_{\frac{1}{2}}$, the infection count of the system drops to 3 ($= \frac{1}{2}I_0$), and never exceeds 3 again, which implies that the virus epidemic has been restricted to a limited area, or equivalently, under control. Further, at $t_0 + \tau_e$, the virus dies out.

Without loss of generality, let $t_0 = 0$, and denote $I(0)$ as I_0 (and $C(0) = C_0$) for the ease of notation. All the events we discuss hereafter take place in window $[0, \tau_e]$.

[2] To be more accurate, r.v. τ_e is almost surely (a.s.) finite, that is, $\mathbb{P}(\tau_e < \infty) = 1$, when the network $\mathcal{G}$ is connected. The reason behind this is that it can be written as the summation of a finite number of exponential r.v.'s, each of which is a.s. finite.

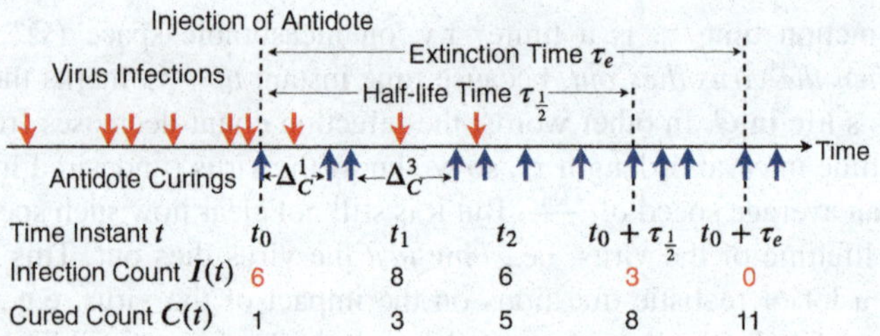

Fig. 2.6 Illustration of the extinction time τ_e and half-life time $\tau_{\frac{1}{2}}$ of the virus under the SIC epidemics example in Fig. 2.5: At time t_0, $C_0 = 1$ copy of antidote is injected into the network

We further assume that both the infection rate β and curing rate γ are constant on every edge in $\mathcal{E}$, following the convention of existing propagation studies.

Under the SIC model, we restate our research question as follows: Consider a pair of conflicting information (x, a_x), in which x is the virus with virulence β, and a_x is the antidote with curing rate γ. At time $t = 0$, C_0 copies of antidote are distributed in network $\mathcal{G}(\mathcal{V}, \mathcal{E})$, when the infection count equals to I_0. Then,

- What is the expected extinction time $\mathbb{E}(\tau_e)$ and half-life time $\mathbb{E}(\tau_{\frac{1}{2}})$ of virus x?
- How to properly select vertices in C_0 to distribute antidotes, such that the lifetime of the virus ($\mathbb{E}(\tau_e)$ and $\mathbb{E}(\tau_{\frac{1}{2}})$) can be shortened?
- Given time t, how to estimate the infection count $I(t)$ and cured count $C(t)$?

We answer these questions sequentially in the following sections. As mentioned in the introduction, the main obstacle of this study resides in the introduction of network $\mathcal{G}$, a structure with numerous topological properties, some of which are of particular importance to the epidemic spreading of information, as evidenced in [16, 21, 30]. Therefore, for the first question, which is also the principle objective of this chapter, we start from simple networks with two special topologies, to gain insights for solutions in networks with arbitrary topologies.

2.3 Lifetime of the Undesired Information in Networks with Simple Topologies

We first examine the lifetime of the virus in the complete network K_n and the star network S_n, which are not only common network topology themselves, but also essential components of complex networked systems.

2.3.1 *Bounds for Complete Networks K_n*

A complete network K_n is essentially a fully connected graph with n vertices, in which for any pair of vertices $v_i, v_j \in V(K_n), i \neq j$, there exists an edge $e(i, j) \in E(K_n)$ between them, and hence $|\mathcal{E}(K_n)| = n(n-1)/2$. This topology is frequently seen in networks that require high reliability, e.g., the network of AS routers, or networks that are densely connected everywhere, e.g., the social network of a household or a community, where every individual is familiar with one another. It has two key characteristics: (1) it is the most densely connected simple graph, with the maximum number of edges; (2) it is regular, as vertices (with the same vertex metrics, e.g., degree and centrality) are inter-changeable. In such networks, the expected lifetime metrics of the undesired information are upper bounded by the following theorem.

Theorem 2.1 *Consider an SIC epidemic in action on a complete network K_n, with curing rate γ, and initial condition (I_0, C_0). The expected extinction time $\mathbb{E}(\tau_e)$ and half-life time $\mathbb{E}(\tau_{\frac{1}{2}})$ of the virus can be bounded above as*

$$\mathbb{E}(\tau_e) < \frac{1}{\gamma n}\left[2 + \ln \frac{(n-1)(n-C_0)}{C_0}\right], \tag{2.6}$$

$$\mathbb{E}(\tau_{\frac{1}{2}}) < \frac{1}{\gamma n}\left[2 + \ln \frac{(n-1)(n-1-\lceil I_0/2\rceil)}{\lceil I_0/2\rceil + 2}\right]; \tag{2.7}$$

when $C_0 \geq 2$, we also have

$$\mathbb{E}(\tau_e) < \frac{2}{\gamma(n-1+C_0)} \ln(n - C_0), \tag{2.8}$$

$$\mathbb{E}(\tau_{\frac{1}{2}}) \leq \frac{2}{\gamma(n-1-\lceil I_0/2\rceil + C_0)}\left(1 + \ln \frac{n - C_0}{\lceil I_0/2\rceil + 1}\right), \tag{2.9}$$

where $n = |\mathcal{V}|$ is the size of the complete network K_n.

To prove this theorem, we first present a set of simple, intuitive bounds on the exact value of the extinction time τ_e and the half-life time $\tau_{\frac{1}{2}}$, as the starting step of the competing case in Theorem 2.1.

Lemma 2.1 *Consider a network $\mathcal{G}(\mathcal{V}, \mathcal{E})$ with $|\mathcal{V}| = n$ under SIC dynamics. At $t = 0$, there are initially I_0 infected nodes and C_0 units of antidote disseminated. We have the following bounds for the extinction time and the half-life time.*

$$\sum_{k=1}^{I_0} \Delta_C^k \leq \tau_e \leq \sum_{k=1}^{n-C_0-1} \Delta_C^k, \tag{2.10}$$

$$\sum_{k=1}^{I_0/2} \Delta_C^k \leq \tau_{\frac{1}{2}} \leq \sum_{k=1}^{n-C_0-1-\lceil I_0/2 \rceil} \Delta_C^k, \tag{2.11}$$

where Δ_C^k *denotes the interval between the* (C_0+k)*-th and* (C_0+k-1)*-th curings in the network, as shown in Fig. 2.6.*

Proof Note that all the initial C_0 curings happen at $t = 0$. Recall our definition of time interval between curings $\Delta_C^k = \tau_C^{k+C_0} - \tau_C^{k+C_0-1}$. At time τ_C^{k-1}, the number of cured vertices can be calculated by $C(\tau_C^{k+C_0-1}) = k - 1 + C_0$.

These upper bounds come from asymmetric immunity, that is, a cured vertex will never be infected again. So the number of cured vertices $C(t)$ is non-decreasing, then the half-life time of the virus $t_{\frac{1}{2}}$ is bounded above by the spreading time of antidote epidemic. At $t_1 = inf\{t > 0 | C(t) = n - \lceil I_0/2 \rceil\}$, the infection count $I(t_1) + S(t_1) = n - C(t_1) = \lceil I_0/2 \rceil$, therefore $I(t) \leq \lceil I_0/2 \rceil, \forall t \geq t_1$.

For lower bounds, τ_e is greater than or equal to the curing time of the initially infected set $\mathcal{I}(0)$, because as the antidote spread, the virus may have collected new victims already. Now consider a 'smart' antidote with the same curing rate γ, or equivalently, let the cured vertices spread the antidote in a 'smart' way, in which it always cures its infected neighbors first. For each realization of the SIC epidemic, this 'smart' antidote distribution is equivalent to re-arranging the sequence of curing events in an SIC epidemic. Similar statement also holds for the half-life time. □

Proof of Theorem 2.1 The process of cured count, $\{C(t)\}_t$ is a continuous time Markov chain with transition rate

$$c \to c+1 \quad \text{at rate} \quad \gamma c(n-c), \quad \forall\, C_0 \leq c \leq n-1,$$

and hence $\mathbb{E}(\Delta_C^c) = \frac{1}{\gamma c(n-c)}$ for any $C_0 \leq c \leq n-1$, because r.v. Δ_C^c is Exponentially distributed with parameter $\gamma c(n-c)$. Then by Lemma 2.1, we have

$$\begin{aligned}\mathbb{E}(\tau_e) \leq \mathbb{E}\Big(\sum_{c=C_0}^{n-1} \Delta_C^c\Big) &= \frac{1}{\gamma n} \sum_{c=C_0}^{n-1} \Big(\frac{1}{c} + \frac{1}{n-c}\Big) \leq \frac{1}{\gamma n}(\mathcal{H}_{n-1} - \mathcal{H}_{C_0-1} + \mathcal{H}_{n-C_0}) \\ &< \frac{1}{\gamma n}\Big[2 + \ln \frac{(n-1)(n-C_0)}{C_0}\Big],\end{aligned} \tag{2.12}$$

where $\mathcal{H}_n$ is the *Harmonic Number*, and $\ln(n+1) < \mathcal{H}_n \leq \ln(n) + 1$. Similarly, the same method can be used to derive upper bounds of half-life time.

$$\mathbb{E}(\tau_{\frac{1}{2}}) \leq \mathbb{E}\left(\sum_{c=C_0}^{n-1-\lceil I_0/2 \rceil} \Delta_C^c\right) \leq \frac{1}{\gamma n}(\mathcal{H}_{n-1-\lceil I_0/2 \rceil} - \mathcal{H}_{\lceil I_0/2 \rceil+1} + \mathcal{H}_{n-1})$$

$$< \frac{1}{\gamma n}\Big[2 + \ln \frac{(n-1)(n-1-\lceil I_0/2 \rceil)}{\lceil I_0/2 \rceil + 2}\Big]. \tag{2.13}$$

Similarly technique can be applied for the case of $C_0 \geq 2$. □

Theorem 2.1 implies that both the expected extinction time and half-life time decrease with size n of the clique as $O(\frac{\log n}{n})$, that is to say, the larger the network, the quicker the undesired information dies, given the same initial condition of the dynamics. The reason behind this phenomenon is that, dense connections among vertices make it difficult for the virus to dodge contact with antidotes. In other words, even though both virus and antidote can spread faster, due to the larger number of edges, dense connections work in favor of the antidote propagation. On the other hand, with respect to the severeness of virus infection upon antidote injection, we can see the half-life time decreases with the initial infection count I_0 as $O(\log \frac{A}{I_0})$, where A is a function of n and C_0 that does not vary with I_0. In this case (larger I_0), the larger gap $\tau_e - \tau_{\frac{1}{2}}$ indicates that, competition between the antidote and the virus is only fierce for a short period of time, and the larger portion of extinction time is spent on extinguishing cornered virus infections.

From the perspective of connectivity, any other simple network topology of the same size n has strictly less edges than the complete graph K_n, as a result of which the undesired information (virus) will die slower with a high probability in these networks, even with the same initial condition (I_0 and C_0). Based on this observation, we also have the following lower bounds as a corollary of Theorem 2.1.

Corollary 2.1 *For an SIC epidemic with curing rate γ on network $\mathcal{G}$ of size n, the expected extinction time and half-life time of the virus can be bounded below as*

$$\mathbb{E}(\tau_e) \geq \frac{1}{\gamma n}\big[\mathcal{H}_{C_0+I_0-1} - \mathcal{H}_{C_0-1} + \mathcal{H}_{n-C_0} - \mathcal{H}_{n-C_0-I_0}\big], \tag{2.14}$$

$$\mathbb{E}(\tau_{\frac{1}{2}}) \geq \frac{1}{\gamma n}\big[\mathcal{H}_{C_0+\lceil I_0/2 \rceil-1} - \mathcal{H}_{C_0-1} + \mathcal{H}_{n-C_0} - \mathcal{H}_{n-C_0-\lceil I_0/2 \rceil}\big], \tag{2.15}$$

where $\mathcal{H}_k = \sum_{j=1}^{k} \frac{1}{j}$ is the k-th Harmonic Number.

Proof Let Δ_C^k denote the time interval between the (C_0+k)-th and the (C_0+k+1)-th curing event, as shown in Fig. 2.6. The expected extinction time and half-life time can be lower bounded by the time that the (C_0+I_0)-th and $(C_0+\lfloor I_0/2 \rfloor)$-th vertex first receives a copy of the antidote respectively, which are shown in Lemma 2.1. On the other hand, the spreading time of the antidote is statistically bounded below by that in a complete network, i.e., when $\mathcal{G} = K_n$. In this case, the value of an inter-curing interval follows an Exponential distribution, that is, $\Delta_C^k \sim Exp(\gamma(C_0+k-1)(n-k-C_0+1))$, because it is the minimum r.v. out of $n-C_0-k$ exponentially distributed r.v.'s. Therefore,

$$\mathbb{E}(\tau_e) \geq \mathbb{E}\left(\sum_{k=1}^{I_0} \Delta_C^k\right) = \sum_{k=1}^{I_0} \mathbb{E}(\Delta_C^k) = \frac{1}{\gamma n}\sum_{k=1}^{I_0}\left[\frac{1}{C_0+k-1} + \frac{1}{n-C_0-k+1}\right], \tag{2.16}$$

and Eq. (2.14) follows from the definition of Harmonic number. Proof of the lower bound of the expected half-life time $\mathbb{E}(\tau_{\frac{1}{2}})$ is similar and hence omitted. □

2.3.2 Bounds for Star Networks S_n

As opposed to the most densely connected complete network K_n, a star network S_n is composed of a hub v_1 and $n-1$ leaves (peripheral vertices), such that every piece of information from a leaf vertex has to go through the hub to reach another vertex. In other words, edges only exist between the hub and leaves, creating a huge bottle-neck in the middle of the network with a high-degree of heterogeneity among vertices. The star topology is also of great importance, due to its natural link to artificial structures, e.g., a WiFi access network within the coverage of an access point, and the ego-network of a high degree node in OSNs.

Consider the same SIC dynamic with infection rate β and curing rate γ on a star S_n. In an extreme case, if a copy of antidote is first given to the hub, then infection count $I(t)$ will be monotonically decreasing with time t, which is less interesting, because it is impossible for the virus to claim new victims, and hence no competition. Therefore, we consider the case that antidotes are distributed to peripheral vertices at time 0 in the following theorem.

Theorem 2.2 *For an SIC epidemic over a star network in which the initial infection count satisfies $I_0 \geq 2$, the expected extinction time $\mathbb{E}(\tau_e)$ and half-life $\mathbb{E}(\tau_{\frac{1}{2}})$ follow upper bounds:*

$$\mathbb{E}(\tau_e) < \frac{1}{\gamma}\left[\frac{1}{C_0} + 1 + \ln(I_0-1) + \frac{\beta}{\gamma C_0(I_0-1)}\right], \tag{2.17}$$

$$\mathbb{E}(\tau_{\frac{1}{2}}) < \frac{1}{\gamma}\left[\frac{1}{C_0} + 1 + \ln(\frac{I_0-1}{\lceil I_0/2\rceil}) + \frac{\beta}{\gamma C_0(I_0-1)}\right]. \tag{2.18}$$

Proof Let r.v. τ_0 denote the time when the hub is cured, then $I(t)$ is non-increasing after τ_0. Let Δ denote the number of new infections until τ_0, T_k denote the time that the cured hub disseminates antidote to the k infected peripheral vertices. Then T_k is the k-th minimum out of the $I_0+\Delta-1$ i.i.d. Exponential r.v.'s with parameter γ. If $k = I_0+\Delta-1$, T_k will be the maximum of k i.i.d. Exponential r.v.'s with parameter γ, or equivalently $\mathbb{E}(T_{I_0+\Delta-1}) = \frac{1}{\gamma}\mathcal{H}_{I_0+\Delta-1}$. Then we can write the expectation of extinction time τ_e, with $\mathbb{E}(\tau_e) = \mathbb{E}(\tau_0)+\mathbb{E}(T_{\Delta+I_0-1})$, where τ_0 satisfies Exponential distribution with parameter $(C_0\gamma)$. Hence, its probability density function $f_{\tau_0}(t) =$

$C_0\gamma e^{-C_0\gamma t}$, and its mean $\mathbb{E}(\tau_0) = \frac{1}{C_0\gamma}$. So

$$\begin{aligned}\mathbb{E}(T_{\Delta+I_0-1}) &= \mathbb{E}\big(\mathbb{E}(T_{\Delta+I_0-1}|\Delta)\big) = \frac{1}{\gamma}\sum_{k=0}^{n-C_0-I_0} \mathcal{H}_{k+I_0-1}\cdot\mathbb{P}(\Delta=k)\\ &< \frac{1}{\gamma}\sum_{k=0}^{n-C_0-I_0}\Big(1+\ln(k+I_0-1)\Big)\cdot\mathbb{P}(\Delta=k). \end{aligned} \tag{2.19}$$

Let function $g(k) = \ln(k + I_0 - 1)$. Notice that $g(\cdot)$ is concave in $[1, n - C_0 - I_0]$. Then

$$\begin{aligned}\mathbb{E}(T_{\Delta+I_0-1}) &< \frac{1}{\gamma} + \frac{1}{\gamma}\mathbb{E}\big(g(\Delta)\big)\\ &= \frac{1}{\gamma} + \frac{1}{\gamma}\mathbb{E}\big(\mathbb{E}(g(\Delta)|\tau_0)\big) = \frac{1}{\gamma} + \frac{1}{\gamma}\int_0^\infty \mathbb{E}\big(g(\Delta)|\tau_0 = t\big)\cdot f_{\tau_0}(t)\mathrm{d}t\\ &\overset{\text{Jensen}}{\leq} \frac{1}{\gamma} + \frac{1}{\gamma}\int_0^\infty g\big(\mathbb{E}(\Delta|\tau_0 = t)\big)\cdot f_{\tau_0}(t)\mathrm{d}t. \end{aligned} \tag{2.20}$$

Now consider r.v. $\mathbb{E}(\Delta|\tau_0 = t)$. Since the network S_n is a star, and Δ is the number of new infections until time instant τ_0, when the hub gets cured, these Δ new infections of different susceptible vertices are mutually independent. Therefore, given a fixed time instant $\tau_0 = t$, r.v. Δ obeys Binomial distribution $B(n - C_0 - I_0, 1 - e^{-\beta t})$, i.e.,

$$\begin{aligned}\mathbb{P}(\Delta = k|\tau_0 = t) &= \mathbb{P}(k \text{ out of } n - C_0 - I_0 \text{ vertices are infected before } t)\\ &= \binom{n - C_0 - I_0}{k}(e^{-\beta t})^{n-C_0-I_0-k}(1 - e^{-\beta t})^k. \end{aligned} \tag{2.21}$$

Consequently, we have the conditional expectation $\mathbb{E}(\Delta|\tau_0 = t) = (n - C_0 - I_0)\cdot(1 - e^{-\beta t})$ as a function of t, which takes value in $[0, \tau_0]$. Note time instant τ_0 is also a r.v., whose range is $[0, \infty)$, so integrating over this range, we have the expected time until the $(\Delta + I_0 - 1)$-th curing as

$$\begin{aligned}\mathbb{E}(T_{\Delta+I_0-1}) &\leq \frac{1}{\gamma} + \frac{1}{\gamma}\int_0^\infty g\big((n - C_0 - I_0)\cdot(1 - e^{-\beta t})\big)\cdot f_{\tau_0}(t)\mathrm{d}t\\ &= \frac{1}{\gamma} - \frac{1}{\gamma}\int_0^\infty \ln\Big[(n - C_0 - 1) - (n - C_0 - I_0)e^{-\beta t}\Big]d\big(e^{-C_0\gamma t}\big)\\ &= \frac{1}{\gamma}\big[1 + \ln(I_0 - 1)\big] + \frac{\beta}{\gamma}\sum_{k=1}^{\infty}\frac{1}{C_0\gamma + k\beta}\cdot\Big(\frac{n - C_0 - I_0}{n - C_0 - 1}\Big)^k\\ &\leq \frac{1}{\gamma}\big[1 + \ln(I_0 - 1)\big] + \frac{\beta}{\gamma^2 C_0(I_0 - 1)}, \end{aligned} \tag{2.22}$$

which is the upper bound of $\mathbb{E}(\tau_e)$ in Eq. (2.17). Similarly, when considering the half-life time, at τ_0, there are $\Delta + I_0 - 1$ infected peripheral vertices. Thus $T_{\Delta+\lceil I_0/2\rceil-1} = t$ indicates the time when there are only $\lceil I_0/2\rceil$ infected vertices left in the peripheral area, or equivalently, $\Delta + \lceil I_0/2\rceil - 1$ out of $\Delta + I_0 - 1$ i.i.d. Exponential r.v's are less than t. Therefore,

$$\begin{aligned}\mathbb{E}(T_{\Delta+\lceil I_0/2\rceil-1}) &= \frac{1}{\gamma}(\mathcal{H}_{\Delta+I_0-1} - \mathcal{H}_{\lceil I_0/2\rceil-1}) \\ &< 1 + \ln(\Delta + I_0 - 1) - \ln(\lceil I_0/2\rceil - 1 + 1) = 1 + \ln\frac{\Delta + I_0 - 1}{\lceil I_0/2\rceil}. \end{aligned} \tag{2.23}$$

Let $g'(k) = \ln\frac{k+I_0-1}{\lceil I_0/2\rceil}$. Then,

$$\begin{aligned}\mathbb{E}(\tau_{\frac{1}{2}}) &= \mathbb{E}(\tau_0) + \mathbb{E}(T_{\Delta+\lceil I_0/2\rceil-1}) \\ &\overset{\text{Jensen}}{<} \frac{1}{C_0\gamma} + \frac{1}{\gamma} + \frac{1}{\gamma}\int_0^{\infty} g'\Big((n - C_0 - I_0)\cdot(1 - e^{-\beta t})\Big)\cdot f_{\tau_0}(t)\mathrm{d}t. \\ &= \frac{1}{\gamma}\left[\frac{1}{C_0} + 1 + \ln(\frac{I_0 - 1}{\lceil I_0/2\rceil}) + \frac{\beta}{\gamma C_0(I_0 - 1)}\right]. \end{aligned} \tag{2.24}$$

So we have the upper bound in Eq. (2.18). □

2.3.3 Numerical Simulation and Discussion

To validate upper bounds (dashed lines) in Theorems 2.1 and 2.2, numerical results (solid lines with markers) with respect to network size n and the initial infection count I_0 are shown in Figs. 2.7 and 2.8, and Figs. 2.9, 2.10, 2.11, and 2.12, respectively.

As can be seen from all these figures, trends of both the expected extinction time and half-life time are well captured by the derived bounds. Comparing results for the two simple topologies, we highlight the following observations.

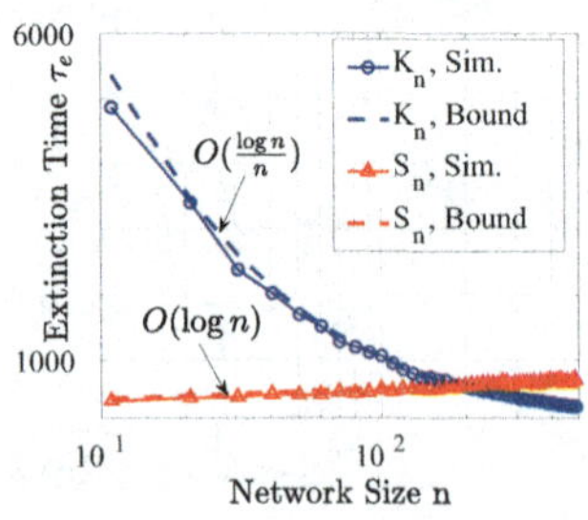

Fig. 2.7 Expected extinction time $\mathbb{E}(\tau_e)$ of an SIC epidemics ($C_0 = 1$, $I_0 = 10$), with respect to the network size n

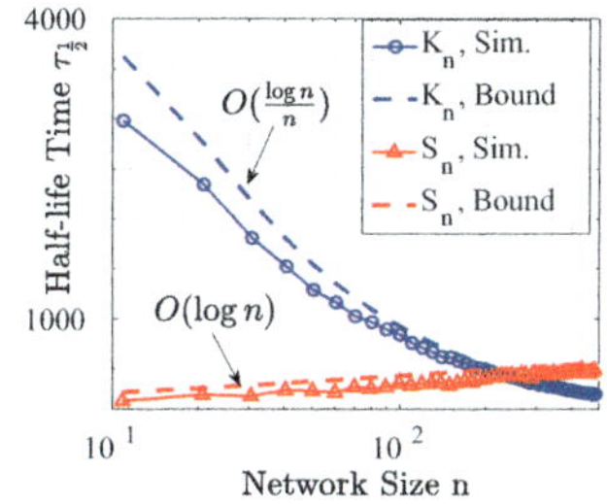

Fig. 2.8 Expected half-life time $\mathbb{E}(\tau_{\frac{1}{2}})$ of an SIC epidemics ($C_0 = 1$, $I_0 = 10$), with respect to the network size n

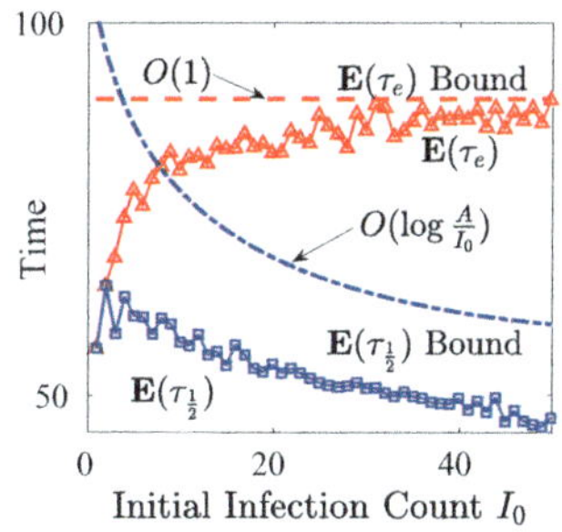

Fig. 2.9 Extinction time $\mathbb{E}(\tau_e)$ and half-life time $\mathbb{E}(\tau_{\frac{1}{2}})$ of the virus in the complete graph K_{100}: $\beta = 0.0005$, $\gamma = 0.001$, $C_0 = 3$

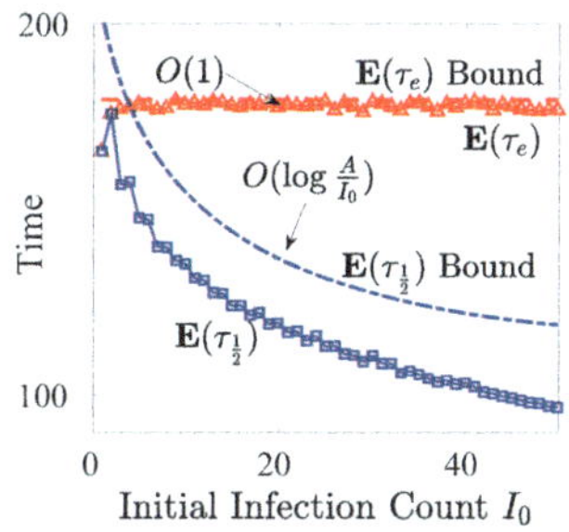

Fig. 2.10 Extinction time $\mathbb{E}(\tau_e)$ and half-life time $\mathbb{E}(\tau_{\frac{1}{2}})$ of the virus in the complete graph K_{100}: $\beta = 0.001$, $\gamma = 0.0005$ $C_0 = 3$

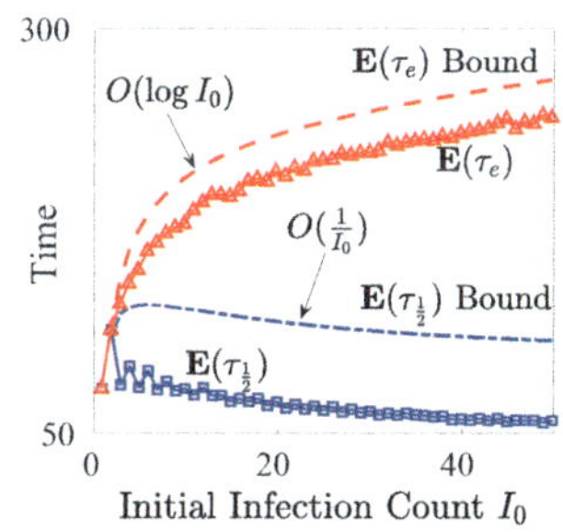

Fig. 2.11 Extinction time $\mathbb{E}(\tau_e)$ and half-life time $\mathbb{E}(\tau_{\frac{1}{2}})$ of the virus in the star network S_{100}: $\beta = 0.01$, $\gamma = 0.02$, $C_0 = 3$

1. There is a clear dichotomy in the lifetime metrics with respect to the network size n. Specifically, virus x dies faster in a larger complete network K_n, but slower in a larger star network S_n, as shown in the semi-log plots of Figs. 2.7 and 2.8. Both $\mathbb{E}(\tau_e)$ and $\mathbb{E}(\tau_{\frac{1}{2}})$ are $O(\frac{\log n}{n})$ for complete networks (decreasing blue lines), while both are $O(\log n)$ for star networks (increasing red lines),

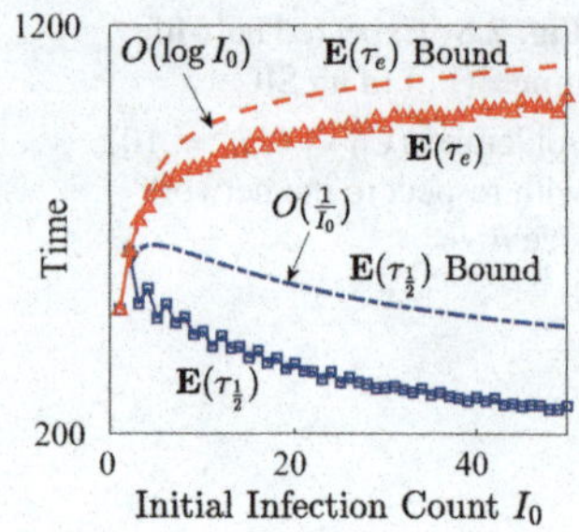

Fig. 2.12 Extinction time $\mathbb{E}(\tau_e)$ and half-life time $\mathbb{E}(\tau_{\frac{1}{2}})$ of the virus in the star network S_{100}: $\beta = 0.01,\ \gamma = 0.005\ C_0 = 3$

as indicated by Theorems 2.1 and 2.2, respectively. This implies an interesting change in the propagation behavior caused by an emerging hub (bottleneck).

2. The infection/curing rates, β and γ, have the same effect on both metrics regardless of network topology, which can be observed by comparing Figs. 2.9 and 2.10, as well as Figs. 2.11 and 2.12. To be more specific, they only lengthen or shorten time intervals between events, but do not change the order of event occurrences, resulting in a homogeneous scaling effect on both lifetime metrics.
3. Impact of the initial infection count I_0 on the lifetime of the virus is shown in Figs. 2.9, 2.10, 2.11, and 2.12, for complete network K_{100} and star network S_{100}, respectively. The mild zig-zag pattern of the half-life time (left end of all blue solid lines with square markers) is caused by rounding-off the threshold $\lfloor \frac{1}{I_0} \rfloor$ when I_0 is small. Though the general trend is the same in both networks, i.e., extinction time increases with I_0, while half-life time decreases with I_0, impact of topology is still apparent, as shown by the different scaling behaviors over I_0.

As two extremes of network topology, the complete graph is regular (every vertex has the same degree) and most densely connected ($|\mathcal{E}| = \frac{n(n-1)}{2}$), while the star graph is highly irregular (due to the existence of the central hub) and sparse ($|\mathcal{E}| = n - 1$). Comparing the two, we observe that, when edges concentrate and form a bottleneck in the network, the impact of network size n and initial condition I_0 changes drastically.

2.4 Lifetime of the Undesired Information in Networks with Arbitrary Topologies

Considering the broad application scenarios of conflicting information propagation, it is likely that the underlying network $\mathcal{G}$ does not have nice topological properties like regularity, and is hence a *complex network* with unique topologies. To study *when* and *how fast* the undesired information dies out in such networks, we examine graph metrics, namely the Cheeger constant $\eta(G)$ and vertex eccentricities $\{\epsilon(v)\}_{v\in\mathcal{V}}$, to quantitatively link topological properties to the lifetime metrics ($\mathbb{E}(\tau_e)$ and $\mathbb{E}(\tau_{\frac{1}{2}})$) of the virus in an SIC epidemics through upper bounds.

2.4.1 Bounds by Considering the Edge-Expansion Property

Recall in the comparison between complete networks and star networks, we find that the key to the change of propagation behavior is the existence of the central hub in star S_n, which forms a bottleneck in the network. Therefore, we first consider the metric that quantitatively measures the level of 'bottleneckness' of a network $\mathcal{G}$, i.e., the Cheeger constant $\eta(\mathcal{G}) := \inf_{S \subset \mathcal{V}, |S| \leq n/2} \frac{|\delta(S)|}{|S|}$, where $\delta(S) := [S, \mathcal{V} \setminus S]$ is the edge-cut of set S, i.e., the set of boundary edges between S and its compliment $\mathcal{V} \setminus S$.

As a graph expansion property [18], $\eta(\mathcal{G})$ identifies the 'narrowest' part of network $\mathcal{G}$, i.e., the minimum boundary edges for as many as vertices. Intuitively, the larger the boundary set, the more difficult it is to break the network into isolated components by disconnecting edges, so $\eta(\mathcal{G})$ and other expansion properties are viewed as indicators of robustness, and hence studied in many applications [18]. It has been shown that $\eta(\mathcal{G})$ is of key importance to the spreading behavior of an epidemic process, under different propagation models, including SI [21, 42], SIS [21], and SIR [16] models. For the SIC model, in which *two* epidemic processes compete in finite time, we show in the following theorem that both the expected extinction time and half-life time are bounded by $\eta(\mathcal{G})$, as $O(\frac{\log n}{\gamma \eta(\mathcal{G})})$.

Theorem 2.3 *For a network $\mathcal{G}$ with Cheeger constant $\eta(\mathcal{G})$, given the initial cured count C_0, the extinction time of the virus in an SIC epidemic with curing rate γ can be bounded above by*

$$\mathbb{E}(\tau_e) \leq \frac{1}{\gamma \eta(\mathcal{G})}[a \ln(n+b) + c]. \tag{2.25}$$

(1) If $C_0 = 1$, $a = \frac{2}{ln 2}$, $b = 0$, $c = 2\gamma_E$;
(2) If $1 < C_0 < n/2$, $a = \frac{2}{\ln 4(C_0 - 1)}$, $b = 0$, and $c = -\left[\frac{4C_0 - 5}{8(C_0 - 1)^2} + \gamma_E\right]$;
(3) If $C_0 \geq n/2$, $a = 1$, $b = -C_0 + 1$, $c = \frac{1}{2(n - C_0 + 1)} + \gamma_E$,

where $\gamma_E \simeq 0.577$ is the Euler-Mascheroni constant. When $2 \leq C_0 < n/2$, the expected half-life time can also be upper bounded by[3]

$$\mathbb{E}(\tau_{\frac{1}{2}}) \leq \frac{1}{\gamma \eta(\mathcal{G})}\left[\ln \frac{n^2}{2(C_0 - 1)I_0} - \frac{4C_0 - 5}{8(C_0 - 1)^2} + \frac{2I_0 + 1}{I_0^2}\right], \tag{2.26}$$

and when $C_0 = 1$, $\mathbb{E}(\tau_{\frac{1}{2}}) \leq \frac{1}{\gamma \eta(\mathcal{G})}\left[\ln \frac{n^2}{2I_0} + \frac{2I_0 + 1}{I_0^2}\right]$.

[3] For the case of $C_0 > n/2$, it is with high probability that $\tau_{\frac{1}{2}} = 0$, which is a less interesting case and hence not discussed.

Proof Let $\mathcal{V}_k := \{V | V \subset \mathcal{V}, |V| = k\}$ be the set of all vertex sets containing k vertices from the network $\mathcal{G}(\mathcal{V}, \mathcal{E})$. Let $C_k \in \mathcal{V}_k$ be the set of cured k vertices when $C(t) = k$. Let $T_k := \inf\{t | C(t) = k\}$ be the time that the cured count $C(t)$ reaches k. Note that $C(t)$ is monotonically increasing, and $C(0) = C_0$, $T_{C_0} = 0$. For any $k \geq C_0$,

$$\mathbb{E}(T_{k+1} - T_k) = \sum_{A \subset \mathcal{V}_k} \mathbb{E}\Big(T_{k+1} - T_k | C(t) = A\Big)\mathbb{P}(C(t) = A\Big) = \mathbb{E}\left(\frac{1}{\gamma |\delta(C_k)|}\right), \tag{2.27}$$

where $\delta(C_k) = [C_k, \mathcal{V} \setminus C_k]$ is the edge cut of set C_k, and

$$|\delta(C_k)| \geq \begin{cases} \eta(\mathcal{G})k, & k < \frac{n}{2}, \\ \eta(\mathcal{G})(n-k), & k \geq \frac{n}{2}. \end{cases}$$

Also, by Lemma 2.1, we have $\mathbb{E}(\tau_e) \leq \sum_{k=C_0}^{n-1} \mathbb{E}(T_{k+1} - T_k)$. So when $2 \leq C_0 < \frac{n}{2}$, the expected extinction time can be upper bounded by

$$\begin{aligned} \mathbb{E}(\tau_e) &\leq \sum_{k=C_0}^{n/2} \frac{1}{\gamma\eta(\mathcal{G})k} + \sum_{k=\frac{n}{2}+1}^{n-1} \frac{1}{\gamma\eta(\mathcal{G})(n-k)} = \frac{1}{\gamma\eta(\mathcal{G})}\left(2\mathcal{H}_{\frac{n}{2}} - \frac{2}{n} - \mathcal{H}_{C_0-1}\right) \\ &\overset{\text{Franel}}{<} \frac{1}{\gamma\eta(\mathcal{G})}\left[\ln\frac{n^2}{4(C_0-1)} - \frac{4C_0-5}{8(C_0-1)^2} - \gamma_E\right], \end{aligned} \tag{2.28}$$

where the last line follows from Franel's Inequality [36]. Similarly, the expected half-life time can be bounded by

$$\mathbb{E}(\tau_{\frac{1}{2}}) \leq \sum_{k=C_0}^{n/2} \frac{1}{\gamma\eta(\mathcal{G})k} + \sum_{k=\frac{n}{2}+1}^{n-1-\lceil I_0/2 \rceil} \frac{1}{\gamma\eta(\mathcal{G})(n-k)}. \tag{2.29}$$

As for $C_0 \geq \frac{n}{2}$, the upper bound can be obtained by considering the bound $\mathbb{E}(\tau_e) \leq \sum_{k=C_0}^{n-1} \frac{1}{\gamma\eta(\mathcal{G})(n-k)}$. □

Though not a sharp bound, Theorem 2.3 confirms the importance of the edge-expansion property, i.e., the *Cheeger* constant $\eta(\mathcal{G})$, in epidemic spreading of information, particularly two pieces of conflicting information studied in this chapter. As can be seen, the more 'expandable' (strong connectivity with less vertices, towards the clique topology), the less time it takes to eliminate the virus, which explains the shorter rumor circulation time [41] than before, as OSNs (and Internet at large) become more connected nowadays. Particularly for the two simple network topologies, clique and star, studied in the previous section, Theorem 2.3 clearly explains the dichotomy of the scaling laws over the network size n: (1)

$\eta(K_n) = n/2 = O(n)$, which implies the $O(\frac{\log n}{n})$ scaling for the complete network K_n; while (2) $\eta(S_n) = O(1)$ for the star network S_n, and hence the $O(\log n)$ increasing over network size n.

For some large networks with special topological properties [18], existing results on their edge expansion properties allow us to estimate the lifetime of the undesired information in such networks as the network grow in size. For instance, Krishnasamy et al. showed that for Erdös-Rényi random graph (ER) $G(n, p)$ with $p > \frac{32 \log n}{4}$, there is a high probability (over $1 - \frac{1}{n^2}$) that $\eta(G) \geq \frac{np}{4}$ [21, Corollary 2]. Then by Theorem 2.3, the expected extinction time is $O(\frac{logn}{\gamma n})$ with high probability.

For a complex network with arbitrary topology, however, obtaining the Cheeger constant is a well-known hard problem (NP-complete, as shown by [27]), especially when the network is large. In this case, $\eta(\mathcal{G})$ can be bounded by the *Cheeger Inequality* [11, 18], $\frac{\lambda_1}{2} \leq \eta(\mathcal{G}) \leq 2\sqrt{\lambda_1}$, where λ_1 is the second smallest eigenvalue of the graph Laplacian[4] of network $\mathcal{G}$. As we will show in the simulation later, $\frac{\lambda_1}{2}$ does not lead to a tight bound of the extinction time, so next we consider properties that are more accessible for general large networks.

2.4.2 Bounds by Considering Vertex Eccentricity

Recall in Theorem 2.2, after the hub of a star network is cured, the extinction time of the virus is bounded by the last curing event of an infected peripheral vertex, which depends on the longest time it takes a copy of antidote to pass though multiple hub-peripheral paths of length 1. In a general network $\mathcal{G}$, if we view every initially cured vertex $v \in C(0)$ as a 'hub', the extinction time also depends on such hub-peripheral paths, whose lengths are measured in hop-count distances $\text{dist}(\cdot, \cdot)$. In practice, distance between two vertices is easy to obtain from the adjacency matrix of a graph, and is therefore widely used in various applications, such as routing and influential node detection. Based on this graph metric, the *eccentricity* of a vertex v in network $\mathcal{G}$ is defined as the longest distance of between v and any other vertex in $\mathcal{V}$, that is, $\epsilon_{\mathcal{G}}(v) := \max_{u \in \mathcal{V}} \text{dist}_{\mathcal{G}}(u, v)$, and the *diameter* of network $\mathcal{G}$ is defined as the largest eccentricity, i.e., $\text{diam}(\mathcal{G}) := \max_{v \in \mathcal{V}} \epsilon_{\mathcal{G}}(v)$.

As the first step, consider the case that a single copy of antidote is distributed to vertex $c \in \mathcal{V}$ at time 0, i.e. $C_0 = \{c\}$. Then propagation time of the antidote to any specific vertex i can be bounded by the following Lemma.

[4] Let $\mathbf{A}$ denote the adjacency matrix of $\mathcal{G}$. The graph Laplacian of $\mathcal{G}$ is defined as $\mathcal{L}(\mathcal{G}) = diag(\mathbf{D}) - \mathbf{A}$, where $\mathbf{D}$ is the degree sequence of $\mathcal{G}$, and $diag(\mathbf{D})$ is the diagonal matrix with $\mathbf{D}$ as its main diagonal. Since $\mathcal{G}$ is undirected and connected, its Laplacian $\mathcal{L}(\mathcal{G})$ is symmetric and positive-semidefinite, and has n non-negative real eigenvalues. Among these, the second smallest eigenvalue λ_1, referred to as the *algebraic connectivity*, measures the expanding property of $\mathcal{G}$. Particularly, the lower bound of $\eta(\mathcal{G})$ is referred to as the Buser Inequality.

Lemma 2.2 *Let $T_{c,i}$ denote the time that vertex i gets a copy of the antidote originated from vertex $c \in C_0$. We have*

$$\mathbb{E}(T_{c,i}) \leq \frac{\text{dist}(c, i)}{\gamma}. \tag{2.30}$$

Proof Let $\{P_k\}_{k=1}^{K}$ denote the set of paths between vertex c and i, such that their lengths are in an ascending order, i.e., $l_k \leq l_{k+1}$. For any $1 \leq k \leq K$, we have

$$\text{dist}_{\mathcal{G}}(c, i) = l_1 \leq l_k \leq l_K. \tag{2.31}$$

Consequently, the curing time $T_{c,i}$ of vertex i (by vertex c) can be re-written as

$$T_{c,i} = \min_{1 \leq k \leq K} T_{c,i}^{k} \leq T_{c,i}^{1}, \tag{2.32}$$

where $T_{c,i}^{k}$ is the attempted curing time of vertex i, by the antidote copy originated from c, and transmitted along path P_k. For every k, time $T_{c,i}^{k}$ is the sum of l_k i.i.d. Exponential r.v.'s with mean $\frac{1}{\gamma}$. Consequently, r.v. $T_{c,i}^{k}$ satisfies Gamma distribution with mean $\mu_k = \frac{l_k}{\gamma}$ and variance $\sigma_k^2 = \frac{l_k}{\gamma^2}$, i.e., $T_{c,i}^{k} \sim \boldsymbol{\Gamma}(l_k, \gamma)$. Then

$$\mathbb{E}(T_{c,i}) = \mathbb{E}(\min_{1 \leq k \leq K} T_{c,i}^{k}) \leq \min_{1 \leq k \leq K} \mathbb{E}(T_{c,i}^{k}) = \frac{l_1}{\gamma} = \frac{\text{dist}_{\mathcal{G}}(c, i)}{\gamma} \tag{2.33}$$

gives us the upper bound in Eq. (2.30). □

Lemma 2.2 provides an upper bound of the expected antidote dissemination time from a designated vertex c to vertex i, which is also an upper bound of the curing time of vertex i, if $c \in C(0)$. Based on this bound, the extinction time of the undesired information can be analyzed with graph augmentation.

Theorem 2.4 *Let $\mathcal{G}_V(\mathcal{V}/V, \mathcal{E}')$ denote the resulting graph induced by contracting vertices in set $V \subset \mathcal{V}$ to a single vertex $f(V)$, and removing all multiple edges*[5]*. Given that antidotes are distributed to the set C_0 at time $t = 0$, The expected extinction time τ_e of the virus can be upper bounded by*

[5] Graph $\mathcal{G}_V$ is the quotient graph of $\mathcal{G}$ through an equivalence relationship induced by partition $\{V, \{v_1\}, \{v_2\}, \cdots\}$ where $v_i \in \mathcal{V} \backslash V$, so that its vertex set $\mathcal{V}/V = (\mathcal{V} \backslash V) \cup f(V)$ has $|\mathcal{V}| - |V| + 1$ vertices.

$$\mathbb{E}(\tau_e) \le \frac{1}{\gamma} \begin{cases} \dfrac{2 \ln K^*_{\mathcal{G}/C_0}}{1-(K^*_{\mathcal{G}/C_0})^{-\frac{1}{\epsilon_{\mathcal{G}_{C_0}}(f(C_0))}}}, & (w.h.p.) \text{ if } \epsilon_{\mathcal{G}_{C_0}}(f(C_0)) \le 10, \\ \epsilon_{\mathcal{G}_{C_0}}(f(C_0)) + \sqrt{(K^s - 1)\epsilon_{\mathcal{G}_{C_0}}(f(C_0))}, & \text{in general,} \end{cases} \tag{2.34}$$

where $K^*_{\mathcal{G}/C_0}$ is the number of longest shortest paths starting from the contracted vertex $f(C_0)$ in the quotient graph $\mathcal{G}_V(\mathcal{V}/V, \mathcal{E}')$, and $K^s \ge K^*_{\mathcal{G}/C_0} + 1$ is the number of shortest paths that start from $f(C_0)$, and are longer than $\epsilon_{\mathcal{G}_{C_0}}(f(C_0)) - s$.

To prove Theorem 2.4, which upper bounds the extinction time with the curing time through dominant paths, we need to upper(lower, respectively) bound the maximum(minimum) of summations of multiple Exponential variables, i.e., multiple r.v.'s with Gamma distribution, to address which we first present a technical lemma.

Lemma 2.3 *Let $\overline{X_n} = \max_{1 \le i \le n} X_i$ denote the maximum of n r.v.'s, and $X_n^* = \min_{1 \le i \le n} X_i$ denote the minimum of them, where each $X_i \sim \boldsymbol{\Gamma}(k_i, \theta_i)$. Then $\mathbb{E}(X_n^*)$ and $\mathbb{E}(\overline{X_n})$ are bounded:*

$$\mathbb{E}(\overline{X_n}) \le \max_i \left\{ \frac{k_i}{\theta_i} \right\} + \left(\frac{n-1}{n} \sum_{i=1}^{n} \frac{k_i}{\theta_i^2} \right)^{1/2}, \tag{2.35}$$

$$\mathbb{E}(X_n^*) \ge \min_i \left\{ \frac{k_i}{\theta_i} \right\} - \left(\frac{n-1}{n} \sum_{i=1}^{n} \frac{k_i}{\theta_i^2} \right)^{1/2}. \tag{2.36}$$

When $k_i = k$ and $\theta_i = \theta$ for every $1 \le i \le n$, such that $\{X_i\}_i$ are a set of i.i.d. Gamma r.v.'s, tighter bounds exist:

$$\mathbb{E}(\overline{X_n}) \le \frac{2 \ln n}{\theta(1 - n^{-1/k})}, \tag{2.37}$$

$$\mathbb{E}(X_n^*) \ge \frac{k \ln 2 - \ln n}{\theta}. \tag{2.38}$$

Proof Equations (2.35) and (2.36) follow from [2, Theorem 2.1] and [2, Corollary 2.1] by plugging in the mean $\mu_i = \mathbb{E}(X_i) = \frac{k_i}{\theta_i}$ and variance $\sigma_i^2 = \frac{k_i}{\theta_i^2}$ of Gamma distributions.

For the i.i.d. case, Eq. (2.37) is proved in [14, Eq. (7)]. Now we prove the lower bound in Eq. (2.38). Let r.v. $Y_i = -X_i$, then $X_n^* = \min_i X_i = -\max_i Y_i \triangleq \overline{Y_n}$. The moment generating function (MGF) of r.v. Y_i can be derived as $M_Y(t) = M_X(-t) = (1 + \frac{t}{\theta})^{-k}$, where $t < \theta$ as required by MGF $M_X(t)$ of r.v. X_i.

Let set $D(M) = \{t \ge 0 | M_Y(t) \ge 1\}$. Then with the technique from [14, Eq. (6)],

$$\mathbb{E}(\overline{Y_n}) \le \inf_{t \in D(M)} \frac{1}{t} [\ln n + \ln M_Y(t)] \le \frac{1}{t} \left[\ln n - k \ln(1 + \frac{t}{\theta}) \right] \triangleq g(t), \tag{2.39}$$

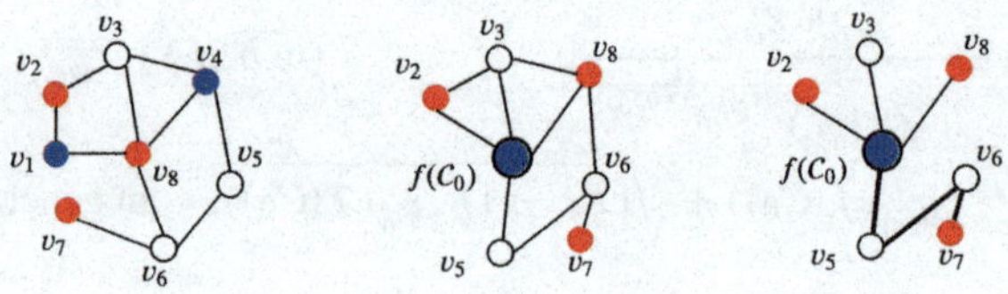

Fig. 2.13 The quotient graph $\mathcal{G}_{C_0}$ after contracting cured set C_0 into vertex $f(C_0)$ in the original graph $\mathcal{G}$, in which $C_0 = \{v_1, v_4\}$. After the contraction, there is $K^*_{\mathcal{G}/C_0} = 1$ path of length $\epsilon_{\mathcal{G}_{C_0}}(f(C_0)) = 3$ from $f(C_0)$ to vertex v_7, and $K' = 1$

which is true for every $t \in [0, \theta)$. Notice that $g(t)$ is monotonically decreasing in $[0, \theta]$, so a tighter bound of $\mathbb{E}(Y_n)$ can be derived as $g(\theta)$. Then Eq. (2.36) follows from the fact that $E(X_n) = -E(Y_n) \geq -g(\theta)$. □

With the preparation of Lemma 2.3, we are now able to prove the upper bound of extinction time in Theorem 2.4.

Proof of Theorem 2.4 For an SIC epidemic on graph $\mathcal{G}$, the extinction time of the virus can be bounded by

$$\max_{i\in\mathcal{I}_0}\left\{\min_{c\in C_0} T_{c,i}\right\} \leq \tau_e \leq \max_{i\in\mathcal{V}\setminus C_0}\left\{\min_{c\in C_0} T_{c,i}\right\}, \tag{2.40}$$

where $T_{c,i}$ is the curing time of vertex i by vertex c.

First we show when $C_0 \geq 2$, the extinction time τ_e on the original graph $\mathcal{G}$ is upper bounded by the extinction time $\widehat{\tau_e}$ of the SIC epidemic on the quotient graph $\mathcal{G}_{C_0}$, in which one unit of antidote is distributed to vertex $f(C_0)$ at time $t = 0$.

An example of the graph contraction procedure is shown in Fig. 2.13. Consider vertex $v_1, v_4 \in C_0$. Evolution of the SIC epidemic (spread of virus and antidote) will not be affected by adding an edge (v_1, v_4) to $\mathcal{G}$, no matter such edge (v_1, v_4) exists in $\mathcal{G}$ or not. This is because: (1) transmission of antidotes between any vertex pair is independent with transmission actions between others; and (2) transmission of antidote between u and v will not result in any state change. Since v_1 and v_4 will stay in *cured* state forever, we can combine and contract them into one vertex $f(\{v_1, v_4\})$ and keep all their edges in $\mathcal{G}$, without affecting the evolution process. The contraction process may result in multiple edges, by removing which the extinction time may increase. Then by induction on the contraction of C_0, the extinction time $\tau_e \leq \widehat{\tau_e}$, which is the extinction time of the virus on the quotient graph $\mathcal{G}_{C_0}$.

Therefore, it is sufficient to consider the SIC epidemic on graph $\mathcal{G}_{C_0}$, with one unit of antidote distributed to $f(C_0)$ at time $t = 0$. Then we have the following inequality,

$$\tau_e \leq \widehat{\tau_e} := \max_{i\in\mathcal{V}\setminus C_0} T_{f(C_0),i} \leq \max_{i\in\mathcal{V}\setminus C_0} T^1_{f(C_0),i}, \tag{2.41}$$

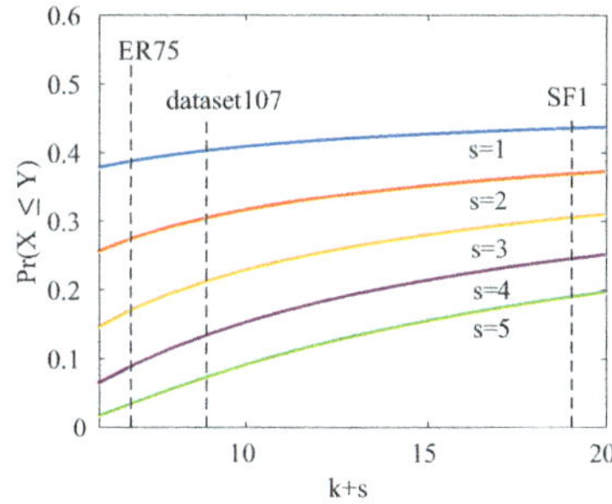

Fig. 2.14 Probability $\mathbb{P}(X \le Y)$ over $k+s$ ($\le$ graph diameter $\text{diam}(\mathcal{G})$), which ranges from 5 to 20

where $T^1_{f(C_0),i} \sim \mathbf{\Gamma}\left(\text{dist}_{\mathcal{G}_{C_0}}(f(C_0), i), \gamma\right)$. The last inequality of Eq. (2.41) follows from Eq. (2.32) in Lemma 2.2. Next we discuss $\max_{i \in \mathcal{V} \setminus C_0} T^1_{f(C_0),i}$ and the eccentricity $\epsilon_{\mathcal{G}_{C_0}}(f(C_0))$ of the central cured hub $f(C_0)$.

By definition, distance between any vertex i and the cured hub $f(C_0)$ satisfies

$$\text{dist}_{\mathcal{G}_{C_0}}(f(C_0), i) \le \epsilon_{\mathcal{G}_{C_0}}(f(C_0)) \le \text{diam}(\mathcal{G}_{C_0}) \le \text{diam}(\mathcal{G}). \tag{2.42}$$

Without loss of generality, suppose $\epsilon_{\mathcal{G}_{C_0}}(f(C_0)) = k + s$ is achieved by the path from $f(C_0)$ to vertex u in the quotient graph $\mathcal{G}_{C_0}$. Consider infected vertex i that is $\text{dist}_{\mathcal{G}_{C_0}}(i, f(C_0)) = k$ hops away from the cured hub $f(C_0)$, such that $s \ge 1$. Let r.v. X and Y denote the time it takes a copy of antidote to reach vertex u and v from the hub $f(C_0)$, respectively. We first show that $\mathbb{P}(X \le Y)$ is small, when $k+s$ is small.

Clearly, r.v. $X \sim \mathbf{\Gamma}(k+s, \theta)$ and $Y \sim \mathbf{\Gamma}(k, \theta)$, so $\frac{X}{X+Y}$ satisfies Beta distribution with parameters $k+s$ and k, and

$$\mathbb{P}(X \le Y) = \mathbb{P}(\frac{X}{X+Y} \le 0.5) = \frac{\Gamma(2k+s)}{\Gamma(k+s)\Gamma(k)} \int_0^{0.5} t^{k+s-1}(1-t)^{k-1} \mathrm{d}t, \tag{2.43}$$

where $\Gamma(k) = \int_0^\infty x^{k-1} e^{-x} dx$ is the Gamma function. It is difficult to bound Eq. (2.43) because of the integral, so we plot $\mathbb{P}(X \le Y)$ for k and s values, such that $k+s$ is in the range of frequently[6] seen network diameters, as shown in Fig. 2.14.

In Fig. 2.14, the dashed vertical lines identify the diameter $\text{diam}(\mathcal{G})$ of three networks used for simulation validation in this section. Among these `dataset107` is a real-world network portion, whose diameter is typical. When $k+s \le \text{diam}(\mathcal{G}) \le 10$, the probability that the antidote will reach u sooner than i, who is s-hops closer than u from $f(C_0)$ is small (<0.2) when $s \ge 3$. In addition, the rate

[6] In most networks we consider (complete graph, star graph, ER graph and scale-free network etc.), the diameter $\text{diam}(\mathcal{G}) = O(\log |\mathcal{V}|)$ is small due to the *small-world* effect, so probability $\mathbb{P}(X \le Y)$ in Eq. (2.43) is also small.

parameter θ of the Gamma distribution equals to the curing rate γ, so when γ is also small, the gap between the two time intervals Y and X is therefore small as well, such that we only need to consider vertices who are located the furthest from vertex $f(C_0)$ in graph $\mathcal{G}_{C_0}$, i.e., the $K^*_{\mathcal{G}/C_0}$ distinct vertices, such as u, satisfying $dist_{\mathcal{G}_{C_0}}(f(C_0), i) = \epsilon_{\mathcal{G}_{C_0}}(f(C_0))$. Then the first term on the right-hand side of Eq. (2.34) follows from the tighter upper bound Eq. (2.37) in Lemma 2.3. Note that when $K^*_{\mathcal{G}/C_0} = 1$, the first term is not well-defined, because the denominator equals to 0. In this case, we take $\mathbb{E}(\tau_e) \leq \frac{\epsilon_{\mathcal{G}_{C_0}}(f(C_0))}{\gamma}$ instead.

If this condition is not satisfied, e.g., when diam($\mathcal{G}$) is large, such that $\mathbb{P}(X \leq Y)$ can not be omitted, we always have the option to consider more paths, which are possibly shorter, to upper bound the extinction time. Then the upper bound (second term in Eq. (2.34)) follows from the upper bound of general distributions, which is illustrated in Eq. (2.35) of Lemma 2.3. □

Note that the eccentricity $\epsilon_{\mathcal{G}_{C_0}}(f(C_0))$ is a property of the augmented quotient graph $\mathcal{G}_{C_0}$. To apply Theorem 2.4, we need to know the exact locations (vertices) where copies of antidotes are disseminated, i.e., set C_0. If such knowledge is not readily available, the extinction time can still be bounded by the diameter (or more generally, distribution of vertex eccentricities) of graph $\mathcal{G}$, as given in the following two corollaries. Proof of Corollaries 2.2 and 2.4 are simple as they directly follow from the fact that $\epsilon_{\mathcal{G}_{C_0}}(f(C_0)) \leq \min_{c \in C_0} \epsilon_{\mathcal{G}}(c) \leq \text{diam}(\mathcal{G})$.

Corollary 2.2 *Particularly when* $C_0 = 1$, *for any initial antidote recipient* $c \in \mathcal{V}$,

$$\mathbb{E}(\tau_e) \leq \frac{1}{\gamma}\left[\text{diam}(\mathcal{G}) + \sqrt{\text{diam}(\mathcal{G})(|\text{Peri}(\mathcal{G})|/2 - 1)}\right], \tag{2.44}$$

where $\text{Peri}(\mathcal{G}) = \{v \in \mathcal{V} \mid \epsilon(v) = \text{diam}(\mathcal{G})\}$ *is the set of peripheral vertices in* $\mathcal{G}$.

Corollary 2.2 is also an upper bound of the expected extinction time $\mathbb{E}(\tau_e)$ for the case of $C_0 > 1$, because the more copies of antidote distributed initially at $t = 0$, the less (statistically bounded) time it takes to fully remove virus from the network, i.e., a shorter extinction time. A direct implication of Corollary 2.2 is that for (often seen) small-world networks, such as social networks, the extinction time is $O(\frac{\log n}{\gamma})$, due to the $O(\log n)$ scaling of the network diameter. Though not a tight bound compared to Theorem 2.3, this corollary is much more accessible in the sense of computation complexity, especially for large networks. Obtaining the eccentricity distribution of a graph, including diameter and peripheral size, is at most $O(n|\mathcal{E}|)$ in time complexity with further speedups [37], which is much faster than obtaining the Cheeger constant $\eta(\mathcal{G})$. By the eccentricity distribution and the initial cured count C_0, a more accurate upper bound can be derived as follows.

Corollary 2.3 *Suppose vertices in* C_0 *are chosen uniformly at random from* $\mathcal{V}$, *and cumulative (probability) density function (CDF) of the eccentricity distribution of graph* $\mathcal{G}$ *is* $F(x)$ *(*$\text{radi}(\mathcal{G}) \leq x \leq \text{diam}(\mathcal{G})$*). Then*

$$\mathbb{E}(\tau_e) \le \frac{1}{\gamma}\left[\mu_{C_0} + \sqrt{\mu_{C_0}\left(\frac{n - C_0}{\text{diam}(\mathcal{G})} - 1\right)}\right], \tag{2.45}$$

where $\mu_{C_0} = \sum_{k=\text{radi}(\mathcal{G})}^{\text{diam}(\mathcal{G})} k[1 - F(k)]^{C_0}$ *is the expected maximum eccentricity of vertices in the cured set* C_0*, and radius* $\text{radi}(\mathcal{G}) := \min_{v \in \mathcal{V}} \epsilon_{\mathcal{G}}(v)$.

Corollary 2.3 stems from Theorem 2.4, by considering the expected maximum eccentricity of the C_0 antidote recipients in C_0, given that antidotes are randomly distributed at time 0. Specifically, for any vertex $v \in V \setminus C_0$, the distance between i and vertex $f(C_0)$ in the quotient graph $\mathcal{G}_{C_0}$ satisfies $dist_{\mathcal{G}_{C_0}}(f(C_0), v) \le \epsilon_{\mathcal{G}_{C_0}}(f(C_0)) \le \min_{c \in C_0} \epsilon_{\mathcal{G}}(c) \le \text{diam}(\mathcal{G})$, which is the diameter of the original graph $\mathcal{G}$. On the other hand, changing perspective to the initial infected set $\mathcal{I}_0$, in which vertices are in infected state at time 0, the expected extinction time $\mathbb{E}(\tau_e)$ can also be bounded below by considering the shortest paths between vertices in set $\mathcal{I}_0$ and set C_0.

Corollary 2.4 *Given the initial infected set* $\mathcal{I}_0 = \mathcal{I}(t = 0)$*, the expected extinction time can be lower bounded by*

$$\mathbb{E}(\tau_e) \ge \frac{1}{\gamma}\left[\text{dist}(C_0, i^*) \ln 2 - ln K_{i^*}\right], \tag{2.46}$$

where $\text{dist}(C_0, i) = \min_{c \in C_0}\{\text{dist}(c, i)\}$, $i^* = \arg\max_{i \in \mathcal{I}_0}\{\text{dist}(C_0, i)\}$, *and* K_{i^*} *is the number of shortest paths*[7] *between set* C_0 *and* i_*.

Proof From Eq. 2.40, we have

$$\tau_e \ge \max_{i \in \mathcal{I}_0}\left\{\min_{c \in C_0} T_{c,i}\right\} \ge \min_{c \in C_0} T_{c,i^*}, \tag{2.47}$$

where $T_{c,i^*} = \min_{1 \le k \le K_{i^*}} T_{c,i^*}^k$ is the minimum of K_{i^*} i.i.d. Gamma r.v.'s, and each $T_{c,i^*}^k \sim \Gamma(\text{dist}(C_0, i), \frac{1}{\gamma})$, which comes from Lemma 2.2. Then the lower bound in Eq. (2.46) follows immediately from Eq. (2.38) in Lemma 2.3. □

2.4.3 Validation in Synthetic and Real-World Networks

To validate the derived bounds, especially the extinction time of the undesired information, we test three networks with distinct topologies, but similar sizes:

---

[7] In practice, obtaining quantity K_{i^*} requires executing path searches repetitively. To avoid high computation load in the simulation, we use the upper bound of K_{i^*}, that is, $|\{i \in \mathcal{I}_0 | \text{dist}(C_0, i) = \text{dist}(C_0, i^*)\}|$ when evaluating the lower bound in Corollary 2.4.

Table 2.1 Statistics of networks used in simulation (Sect. 2.4.3)

Property	Description	ER75	SF1	`dataset107`
n	Network size $\lvert\mathcal{V}\rvert$	1000	1000	1033
$\bar{d}$	Average degree	7.362	1.9980	51.7851
$\text{diam}(\mathcal{G})$	Network diameter, $\max_{v\in\mathcal{V}} \epsilon_{\mathcal{G}}(v)$	7	19	9
$\text{radi}(\mathcal{G})$	Network radius, $\min_{v\in\mathcal{V}} \epsilon_{\mathcal{G}}(v)$	5	10	5
λ_1	Algebraic connectivity	0.7618	0.002	0.1329
$\lvert\text{Peri}(\mathcal{G})\rvert$	Periphery size, $\lvert\{v \in \mathcal{V} \vert \epsilon_{\mathcal{G}}(v) = \text{diam}(\mathcal{G})\}\rvert$	1	8	1

random graph (ER graph ER75 with edge connecting probability $p = 0.0075$, scale-free graph SF1 generated with the Barabási–Albert model by attaching one edge for one vertex each time, and a fraction of Facebook friendship network `dataset107` [25]. Statistics of the three connected graphs, ER75, SF1 and `dataset107` are shown in Table 2.1. Though all of the three networks have around 1000 vertices, the SF1 network has a large diameter ($\text{diam}(\mathcal{G}) > 10$, which indicates that the condition in the first case of Theorem 2.4 does not hold.) and is less connected, compared to ER75 and `dataset107`. This can be observed by the low average degree and algebraic connectivity λ_1. ER75 is more clustered and even, in the sense that it achieves a high algebraic connectivity λ_1 with a relatively low average degree.

Figures 2.15, 2.16, and 2.17 show the simulated extinction time (grey dots) in the three networks, respectively, whose mean $\mathbb{E}(\tau_e)$ are identified with blue '×' markers. Each instance in the x-axis corresponds to the same initial cured set C_0 and infected set $\mathcal{I}_0$, and each instance is repeated 100 times to obtain $\mathbb{E}(\tau_e)$. The red round markers and black squares correspond to the upper bounds in Theorem 2.4 and Corollary 2.4, respectively, both of which rely on vertex eccentricity properties of the network. The red dashed line and green dashed line with triangle markers correspond to the upper bounds presented in Corollary 2.2 that relies on network diameter $\text{diam}(\mathcal{G})$, and Theorem 2.3 that relies on the Cheeger constant[8] $\eta(\mathcal{G})$, respectively.

We highlight the following observations:

1. *Network Topology*: Extinction time $\mathbb{E}(\tau_e)$ varies more violently in SF1 and `dataset107` networks that are less 'regular' (less like clique topology) in the sense that vertices differ more in degree, centrality, etc. (indicating different importance/influential in status of individuals). Therefore, in SF1, due to the larger curing rate γ, and the larger diameter $\text{diam}(\mathcal{G})$, the probability in Eq. (2.43) will not be small enough to be neglected, so the upper bound specifically

[8] Approximating the Cheeger constant $\eta(\mathcal{G})$ with the lower bound $\frac{\lambda_1}{2}$ results in a very loose bound of $\mathbb{E}(\tau_e)$ at $\sim 10^4$, so it is not shown for a better illustration and comparison among other bounds. The green dashed line corresponds to the upper bound in Cheeger's Inequality $\eta(\mathcal{G}) = 2\sqrt{\lambda_1}$, which is presented to illustrate how close Theorem 2.3 can be, if the Cheeger constant reaches its maximum possible value. Hence the star sign in the legend.

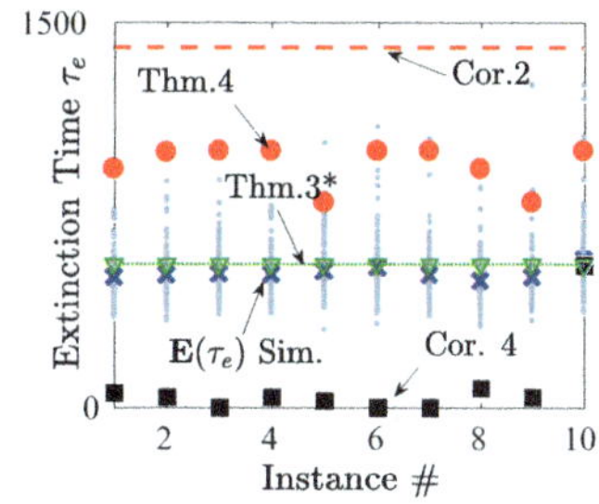

Fig. 2.15 Bounds and simulation of the extinction time $\mathbb{E}(\tau_e)$ in general networks: Erdös-Rényi random graph (ER) network ER75 with $C_0 = 3$, $\gamma = 0.005$

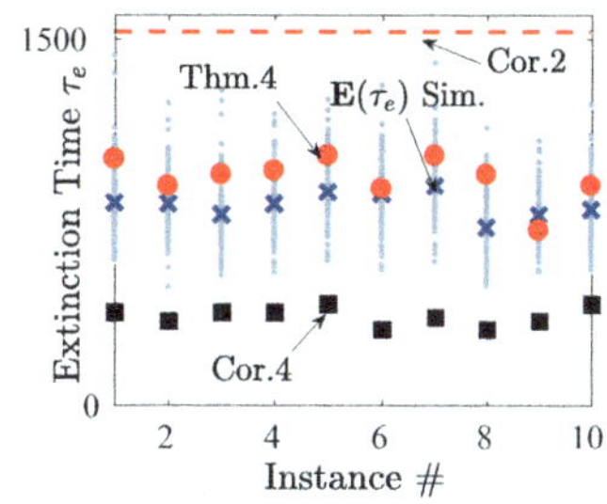

Fig. 2.16 Bounds and simulation of the extinction time $\mathbb{E}(\tau_e)$ in general networks: Scale-free network SF1 with $C_0 = 3$, $\gamma = 0.02$

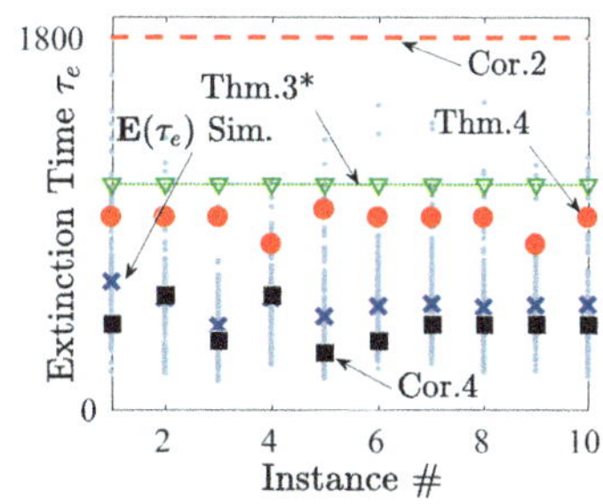

Fig. 2.17 Bounds and simulation of the extinction time $\mathbb{E}(\tau_e)$ in general networks: Real-world network portion `dataset107` with $C_0 = 5$, $\gamma = 0.005$

for i.i.d. Gamma distributions (first term in the right-hand side of Eq. (2.34)) in Theorem 2.4 does not hold any more, but the second term for general distributions still holds, which is adopted as the red round markers in Fig. 2.16. For the 'regular' ER75 and `dataset107` networks, despite their differences in average degree, and algebraic connectivity, their similarity in the vertex eccentricity properties (diameter, radius, and periphery size in Table 2.1) leads to very close extinction time, which indicates the usefulness of Corollary 2.2.

2. *Tightness of the Bounds*: Theorem 2.4 and Corollary 2.4 are tighter in the 'irregular' SF1 and `dataset107`, due to the existence of few 'dominant' longest shortest paths, i.e., K^* is small. In contrast, for the ER75 network, vertex degrees and pair-wise distances are highly homogeneous (small variance in degree distribution), resulting in a large K^* that affects the tightness of both upper and lower bounds. Corollary 2.2 (red dashed line in Figs. 2.15, 2.16, and 2.17 that relates to the network diameter diam($\mathcal{G}$) is an upper bound not only good for the mean, but also for every simulation runs. Theorem 2.3 is not a tight bound. As indicated by the green dashed line with triangle markers, which approximates

the Cheeger constant with its maximum possible value $\eta(G) = 2\sqrt{\lambda_1}$, even so the derived upper bound is not as close as Theorem 2.4, especially in SF1 and `dataset107`.

3. *Implications*: From the proof of Theorem 2.4, another implication is that, the extinction time of the undesired information can be bounded by the number and the length of shortest paths, which go through vertices of the initial cured set C_0. This observation sheds light on antidote distribution (injection of desired information), when used as a counter-measure against the epidemic spreading of undesired information: A general guideline on selecting vertices for set C_0 is to choose vertices that sit on the most number of shortest paths, i.e., vertices with high *betweenness centrality*. To better understand the structural change due to C_0 and its impact to the extinction time of the virus, we discuss antidote distribution in the next section in details.

2.5 Divide-and-Conquer: Leveraging Topology to Control Undesired Information

Desired information can be purposely injected into networked systems to *control* the epidemic spreading of undesired information. For instance, security patches can be distributed to computer users to control the spread of computer malwares.

On the control of undesired epidemics, Preciado et al. developed a convex framework [35] to evaluate the optimal allocation of edge control, immunization and non-replicable antidotes resources, in which the network is modeled as a directed graph. Borgs et al. studied the optimal distribution of non-replicable antidote [5] given that the curing rate, proportional to the units of given antidote, is non-uniform. As for replicable antidote, Khouzani et al. formulated the control strategy [19] with both replicable and non-replicable antidote into an optimal resource allocation problem. Chen utilized optimal control theory to determine the optimal distribution time [9] of a replicable antidote for timely control. However in [5, 9, 19], the information dynamic in the system is described by nonlinear differential equations, which indicates it is a *population dynamic*, rather than a *network dynamic*, or equivalently, topology of the network is not taken into consideration. On the other hand, the influence of network topology is studied in terms of epidemic threshold [23] under immunization and spreading/extinction time [21] under SI or SIS epidemic propagation models, while the dissemination of replicable antidote is not incorporated. To bridge this gap, we explore the distribution strategies of *replicable antidote*, leveraging *topology* in order to control the undesired information.

Apparently, both the extinction time and the half-life time changes with the initial distribution choice, i.e., $C(0)$, so a carefully chosen set of vertices can effectively shorten the lifetime of the virus, as illustrated by a simple example in Fig. 2.18. After the initial distribution, the potential hazard zone (pink shaded region) in the random distribution case is larger than that of the targeted distribution. During the

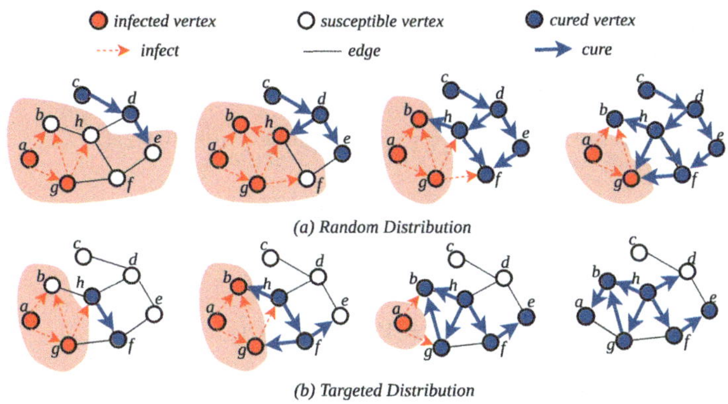

Fig. 2.18 An example of (**a**) random distribution; (**b**) targeted distribution

propagation process, the effect of the replicable antidote is twofold: on individual bases, it cures infected vertices, reducing the infection count; on the other hand, the expanding cured set composes a structure to retain the potential hazard zone of the virus. In the latter case, the influence of network topology is more evident, because the dynamics is changing the topological property of the system.

2.5.1 Topology-Based Antidote Distribution

As shown in the targeted distribution of Fig. 2.18, the potential hazard zone of the virus is restricted to a limited region of the network. In other words, a non-empty set of susceptible vertices are 'quarantined' by the initial antidote distribution, and they will never be infected during the dynamics. We are interested in such 'locking' condition, which effectively restrains the virus propagation. To characterize this effect, we introduce the *initial locking time*, defined as follows.

Definition 2.3 Let $\mathcal{G}^*(t)$ be the induced subgraph of $\mathcal{G}$ by removing the cured vertices $C(t)$. We write $\mathcal{G}^*(t) = \cup_{1 \le i \le k(t)} G_i$, where $G_i(V_i, E_i)$ are components of $\mathcal{G}^*(t)$ and $k(t)$ denote the number of components at time t. The **initial locking time** τ_0 is defined as the first time that $\mathcal{G}^*(t)$ becomes disconnected, or equivalently

$$\tau_0 := \sup\{t > 0 \mid k(t) \ge 2\}. \tag{2.48}$$

Time instant τ_0 marks a critical point of the virus epidemic, because a topological property of the remaining graph $\mathcal{G}^*(t)$, i.e., the *connectivity* of $\mathcal{G}^*(t)$, changes at τ_0. When $t < \tau_0$, $k(t) = 1$, the virus can potentially spread to every corner of the remaining graph $\mathcal{G}^*(t)$. As time t goes beyond τ_0, fragmentation starts to happen in each $G_i \subset \mathcal{G}^*(t)$, and potential hazard of the virus can be treated as under control.

Then we have the following bound of the extinction time $\mathbb{E}(\tau_e)$ over $\mathbb{E}(\tau_0)$.

Theorem 2.5 *Let $C_0 = |C(0)|$. The expected extinction time can be upper bounded by*

$$\mathbb{E}(\tau_e) \leq \mathbb{E}(\tau_0) + \frac{1}{\gamma}[\frac{2}{(n - C_0)\eta(\mathcal{G})} + \max_{1 \leq i \leq k(\tau_0)} \{\text{diam}(G_i)\}], \tag{2.49}$$

where diam(G) *and* $\eta(G)$ *are the diameter and Cheeger constant of graph* G.

Proof Consider r.v. Z_i and Y_i, defined as follows.

$$Z_i := \inf_{t > \tau_0} \{t - \tau_0 \mid |V_i \cap \mathcal{I}(t)| = 1\}, \tag{2.50}$$

$$Y_i := \inf_{t > Z_i + \tau_0} \{t - (Z_i + \tau_0) \mid |V_i \cap \mathcal{I}(t)| = 0\}. \tag{2.51}$$

The physical meaning of Z_i is the time interval between the initial locking and the first vertex in G_i is cured, or equivalently, the minimum of $\delta(G_i)$ i.i.d. exponential r.v.'s with parameter γ, where $\delta(G_i)$ denote the number of edges in edge cut $[V_i, \mathcal{V} \setminus V_i]$. And r.v. Y_i denotes the time interval between $\tau_0 + Z_i$ and the time that all the infected vertices in G_i are cured. Now we re-write $\mathbb{E}(\tau_e)$ with the help of $\mathbb{E}(\tau_0)$ and these r.v.'s,

$$\begin{aligned}
\mathbb{E}(\tau_e) &\leq \mathbb{E}(\tau_0) + \mathbb{E}(\max_{1 \leq i \leq k(\tau_0)} \{Z_i + Y_i\}) \\
&\overset{\text{Jensen}}{\leq} \mathbb{E}(\tau_0) + \max_{1 \leq i \leq k(\tau_0)} \{\mathbb{E}(Z_i) + \mathbb{E}(Y_i)\} \\
&\leq \mathbb{E}(\tau_0) + \max_{1 \leq i \leq k(\tau_0)} \{\mathbb{E}(Z_i)\} + \max_{1 \leq i \leq k(\tau_0)} \{\mathbb{E}(Y_i)\}).
\end{aligned} \tag{2.52}$$

Note that the first inequality of Eq. (2.52) follows from the fact that among all the components of $\mathcal{G}^*(\tau_0)$, some may not contain infected vertices. As the minimum of multiple exponential r.v.'s, $Z_i \sim \text{Exp}(\delta(G_i)\gamma)$. Therefore,

$$\begin{aligned}
\max_{1 \leq i \leq k(\tau_0)} \{\mathbb{E}(Z_i)\} &= \frac{1}{\gamma}[\max_{1 \leq i \leq k(\tau_0)} \{\frac{1}{|\delta(G_i)|}\}] \\
&\leq \frac{1}{\gamma}[\max_{1 \leq i \leq k(\tau_0)} \{\frac{1}{\eta(\mathcal{G}) \min\{|V_i|, n - |V_i|\}}\}] \\
&\leq \frac{1}{\gamma} \frac{2}{(n - C_0)\eta(\mathcal{G})}.
\end{aligned} \tag{2.53}$$

Suppose the first cured vertex in G_i is $v_i \in V_i$, then Y_i is bounded above by the spreading time of the antidote along the shortest-path spanning tree of G_i rooted at v_i. Since $v_i \in {}_i$ can be any vertex that is directed connected to $C(\tau_0)$. Then,

$$Y_i \leq \max_{v_i \in E_i} \{ \sum_{s=1}^{\text{depth}(G_i, v_i)} \Delta_C^s \} \leq \sum_{s=1}^{\text{diam}(G_i)} \Delta_C^s, \tag{2.54}$$

where $\Delta_C^s \sim \text{Exp}(\gamma)$ denote the time intervals between the $(s-1)$-th and the s-th curing, $\text{depth}(G_i, v_i)$ denotes the depth of the shortest-path spanning tree of G_i, and $\text{diam}(G_i)$ is the diameter (length of the longest shortest-path) of G_i. The second inequality of Eq. (2.54) follows from the fact that $\text{depth}(G_i, v_i) \leq \text{diam}(G_i)$, $\forall v_i \in E_i$, so $\sum_{s=1}^{\text{depth}(G_i,v_i)} \Delta_C^s \leq \sum_{s=1}^{\text{diam}(G_i)} \Delta_C^s$, for any $v_i \in E_i$. Therefore,

$$\max_{1\leq i\leq k(\tau_0)} \{\mathbb{E}(Y_i)\} \leq \frac{1}{\gamma}[\max_{1\leq i\leq k(\tau_0)} \{\text{diam}(G_i)\}]. \tag{2.55}$$

Combining Eqs. (2.52), (2.53), and (2.55) completes the proof. □

Theorem 2.5 implies that, the virus infections are defeated by a *divide-and-conquer* strategy. Specifically, the extinction time can be mainly determined by $\mathbb{E}(\tau_0)$ and quantity $max_{1\leq i\leq k(\tau_0)}\{\text{diam}(G_i)\}$, since $\eta(G) \geq \frac{n}{2}$ (the case that all vertices form a single line) indicates the second term in Eq. (2.49) is at most $O(1)$.

2.5.2 *Ideal Antidote Distribution Policy*

From a holistic view, there are two influential factors on the extinction time: the initial antidote distribution policy, and the network topology. The former includes the initial cured count C_0 and the recipients of the C_0 antidotes, i.e. assignment of $C(0)$. Now consider the initial locking time τ_0. Apparently it is decreasing as the initial cured count C_0 increases, as well as $\text{diam}(G_i)$. However, it is not reasonable nor realistic to let C_0 approach n. Consequently, our objective is to find an effective way to distribute as less antidote as possible, on condition that the extinction time can be shortened by the most amount of time. So in this subsection, we assume that there are just enough amounts of antidote, such that $\tau_0 = 0$, i.e., $C(0)$ is a *vertex cut* of $\mathcal{G}$. To understand the effectiveness of the antidote dissemination policy, we define the following metric to quantify the impact of the graph topology on the extinction time with respect to an assignment of $C(0)$.

Definition 2.4 The **hazard index** $\phi(C(0))$ is defined as the maximum diameter of components of the initially disconnected graph $\mathcal{G}^*(t) = \cup_{1\leq i\leq k(t)} G_i$, that is,

$$\phi(C(0)) := \max\left\{\text{diam}(G_i) \mid G_i \subset \mathcal{G}^*(\tau_0), 1 \leq i \leq k(\tau_0)\right\}. \tag{2.56}$$

Further, we provide the following upper bound of $\phi(C(0))$ in terms of $|V_i|$, that is, the number of vertices in each components of $\mathcal{G}^*(0)$.

Theorem 2.6 *Let $d'(\mathcal{G})$ denote the minimum degree of graph $\mathcal{G}$, then*

$$\phi(C(0)) \leq \frac{3}{d'(\mathcal{G})+1} \max_{1\leq i\leq k(\tau_0)} \{|V_i|\} - 1. \tag{2.57}$$

Proof Denote the minimum degree of each component G_i as d'_i. If it satisfies $d'_i \le \frac{|V_i|-2}{2}$[9], then by [7], we have

$$\operatorname{diam}(G_i) \le 3\lfloor \frac{|V_i|}{d'_i+1} \rfloor - \begin{cases} 3, & n \mod (d'_i+1) = 0; \\ 2, & n \mod (d'_i+1) = 1; \\ 1, & \text{otherwise.} \end{cases} \tag{2.58}$$

Further,

$$\begin{aligned} \phi(C(0)) &\le \max_{1\le i\le k(\tau_0)} \{3\lfloor \frac{|V_i|}{d'_i+1} \rfloor - 1\} \le \max_{1\le i\le k(\tau_0)} \{3(\frac{|V_i|}{d'(\mathcal{G}^*(\tau_0))+1}) - 1\} \\ &\overset{*}{\le} \max_{1\le i\le k(\tau_0)} \{3(\frac{|V_i|}{d'(\mathcal{G})+1}) - 1\} \le \frac{3}{d'(\mathcal{G})+1} \max_{1\le i\le k(\tau_0)} \{|V_i|\} - 1. \end{aligned} \tag{2.59}$$

Note that the third inequality in Eq. (2.59) (marked by $*$) is not strict, but still reasonable. Since $d'(\mathcal{G}^*(\tau_0)) \ge d'(\mathcal{G})$ holds except the case that the vertex with the minimal degree (denote as h) is directly adjacent to a vertex in $C(0)$. If the degree of vertex h is decreased much due to the removal of $C(0)$, then it would be more convenient to include h in $C(0)$ at the initial antidote distribution, and in this case, the degree of the remaining vertices is decreased at most 1, due to the fact that $\mathcal{G}$ is simple, that is, $\mathcal{G}$ does not have repetitive edges. □

Theorem 2.6 provides an upper bound of $\phi(C(0))$ in terms of the size of the giant component $\max_{1\le i\le k(\tau_0)}\{|V_i|\}$ in $\mathcal{G}(\tau_0)$, where $\tau_0 = 0$ in this case. In discrete mathematics, this is equivalent to the *most balanced cut problem*, that is, finding such a vertex-cut constrained on different balance requirements, which in our case, is the number of vertices in each component G_i.

On the other hand, clearly from Eq. (2.49), our goal of assigning $C(0)$ is to minimize the hazard index, i.e., $\min_{C(0)}\{\phi(C(0))\}$. From the Moore Bound [28] Inequality, we know that the diameter of each subgraph G_i can be lower bounded by[10]

$$\operatorname{diam}(G_i) \ge \begin{cases} \frac{|V_i|-2}{2}, & d_i = 2, \\ \log_{d_i-1}[(|V_i|-1)\frac{d_i-2}{d_i}) + 1], & d_i > 2, \end{cases} \tag{2.60}$$

where d_i is the maximum degree and $|V_i|$ is the number of vertices of graph G_i respectively. Let $f(d_i, |V_i|) := \log_{d_i-1}[(|V_i|-1)\frac{d_i-2}{d_i}) + 1]$. It is clear that

[9] It is highly unlikely that $d'_i > \frac{|V_i|-2}{2}$, as that would suggest every component G_i is dense.

[10] We remark that the Moore Bound is fairly difficult to obtain, but it is a valid lower bound with respect to the number of vertices of each component.

$f(d_i, |V_i|)$ is decreasing in d_i, but increasing in $|V_i|$. In the mean time, $\frac{|V_i|-2}{2}$ is also increasing in $|V_i|$, irrelevant to d_i. Consequently,

$$f(d_i, |V_i|) \geq f(d(\mathcal{G}), |V_i|), \tag{2.61}$$

where $d(\mathcal{G})$ denotes the maximum degree of the network $\mathcal{G}$.

Now minimizing $\max\{\text{diam}(G_i)\}$ becomes minimizing $\max\{|V_i|\}$, so our goal becomes finding a minimum vertex cut $C(0) \subset \mathcal{V}$, such that in the induced subgraph $\mathcal{G}^*(\tau_0) = \mathcal{G}\backslash C(0)$, $\min_{C(0)}\{\max_i\{|V_i|\}\}$ can be achieved. Again, the upper bound of $\phi(C(0))$, (and hence that of $\mathbb{E}(\tau_e)$) is relevant to the maximum number of vertices in each component of $\mathcal{G}^*(\tau_0)$, which is actually the *minimum most balanced vertex-cut problem*. By assigning $C(0)$ to this cut, the upper bound of $\mathbb{E}(\tau_e)$ in Eq. (2.49) can be tightened, because $\max_i\{\text{diam}(G_i)\}$ is reduced.

Therefore, to maximize the effect of antidote that can propagate itself, as well as to minimize the extinction time, an ideal initial antidote distribution strategy is to assign $C(0)$ to the minimum most balanced vertex-cut of the network $\mathcal{G}$.

2.5.3 *Practical Approaches*

In real-world networks that are complex in topology, nonetheless, it is impractical to find a small vertex-cut, let alone a minimum most balanced vertex-cut. For example, in a complete graph K_n, the minimum vertex-cut contains $n-1$ vertices. It translates to either a long τ_0, or a large enough C_0, but $\mathbb{E}(Y_i) \simeq \frac{1}{\gamma(n-1)}$, in which case the upper bounds in the previous sections are more applicable. Moreover, searching for the minimum most balanced cut or directly searching for vertices, whose removal will result in a smaller $\phi(C(0))$ is quite difficult. In fact, finding the minimum most balanced cut for general graphs is NP-Hard [20]. So we introduce the following applicable and practical approaches under the guideline of minimizing $\phi(C(0))$, or minimizing the size of the giant component in $\mathcal{G}^*(\tau_0)$.

1. *Betweenness centrality* of a vertex v is defined as $g(v) := \sum_{s\neq v\neq t} \frac{\delta_{st}(v)}{\delta_{st}}$, where δ_{st} denotes the number of shortest paths between vertex s and v, while $\delta_{st}(v)$ are the number of those paths that passes through vertex v. Parameter $g(v)$ indicates how likely vertex v sits in other vertices' shortest paths. Removal of a vertex with high g value will likely break more shortest paths, which is an effective measure as indicated by the extinction time bound in the previous section. However, it requires global knowledge and takes $O(|\mathcal{V}||\mathcal{E}|)$ time [6] to calculate.
2. *Clustering Coefficient* of a vertex v is defined as $C(v) := \frac{2|\{e_{st}|e_{st}, e_{vs}, e_{vt} \in \mathbb{E}\}|}{d(v)[d(v)-1]}$, which shows how densely connected is v's neighborhood. To calculate $C(v)$, knowledge of vertices within the distance of two to v is required. The reason of using C value as an indicator is that when $C(v)$ is a small but strictly positive value, it implies that v's neighbor is not well-connected, and relies v to function

as a bridge between its neighbors. This is especially true when the graph does not have many 'long edges'.

3. *Degree* $d(v)$ of vertex v is easy to attain, because it only requires knowledge of one-hop neighbors. Higher degree indicates a vertex has a higher chance of being a hub. Hence removing such a vertex will disconnect a lot of edges.

As a practical simple approach, we first calculate the g, C and d values for each vertex, then sort them to find the best candidates. Considering the sorting process requires global knowledge of the graph, it could be substituted with a cut-off mechanism with predetermined threshold values in real world implementations.

2.5.4 Numerical Results and Discussion

To examine the effectiveness of the proposed antidote dissemination policy, we conduct simulation of SIC dynamics on two network fractions from Facebook [25] (now "Meta"), identified as `dataset0`[11] and `dataset348`. Statistics of the two network fractions are listed in Table 2.2. Their topologies are illustrated in Fig. 2.19, where color of each vertex indicates the betweenness centrality value of the vertex (blue for low g, and red for high g). Intuitively from the figure, we can tell that `dataset0` is more 'scattered' than `dataset348`. This implies that it will be easier (with a smaller $C_0 = |C(0)|$) to achieve the locking condition in `dataset0`.

Applying different initial antidote distribution strategies, i.e., different assignments of $C(0)$ on `dataset0` and `dataset348`, the resulting topological changes can be observed in Fig. 2.20, which can be used to predict the effectiveness of these strategies through the lens of $\mathbb{E}(\tau_e)$ and $\mathbb{E}(\tau_{\frac{1}{2}})$. It is quite interesting to see that betweenness centrality-based approach (tagged '$g(v)$', red lines with round markers) outperforms by clustering coefficient-based approach (tagged '$C(v)$', green lines with square markers) in `dataset348` (dotted lines), while in `dataset0` (dashed lines) it is the opposite. The possible reason is that `dataset348` is much denser (avg. degree 57) and more clustered (avg. ccfs 0.544) than `dataset0`,

Table 2.2 Statistics of the two networks: `dataset0` and `dataset348`

Statistics	Description	`dataset0`	`dataset348`
$n = \|\mathcal{V}\|$	Order (number of vertices)	324	224
$\|\mathcal{E}\|$	Size (number of edges)	5028	6384
$\bar{d} = 2\|\mathcal{E}\|/n$	Average degree	31.037	57
diam($\mathcal{G}$)	Network diameter	11	9
$\bar{C}$	Average ccfs	0.522	0.544
$\bar{l}$	Average path length	3.573	3.042

[11] Originally, `dataset0` contains 342 vertices and is disconnected, so we select the giant component to be dataset0 for this simulation validation.

Fig. 2.19 Topology of `dataset0` network and `dataset348` network, respectively

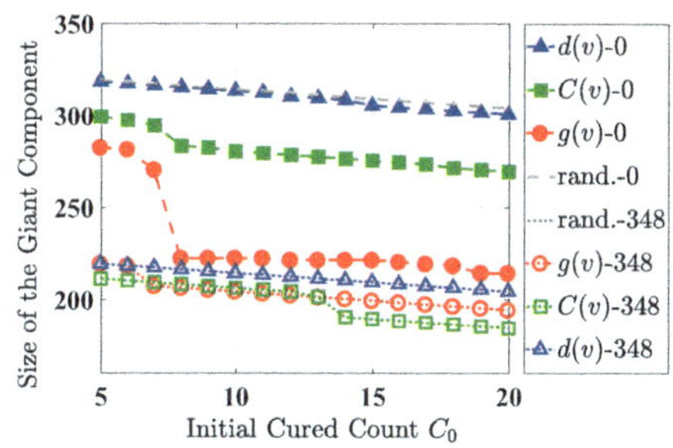

Fig. 2.20 The change of the giant component size over time in `dataset0` and `dataset348` networks

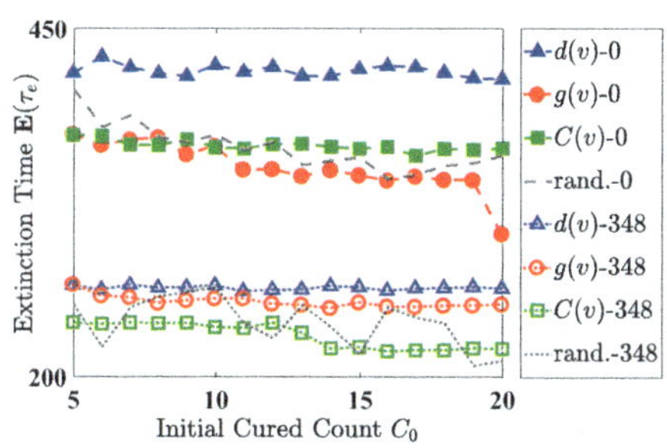

Fig. 2.21 Extinction time $\mathbb{E}(\tau_e)$ on the two networks

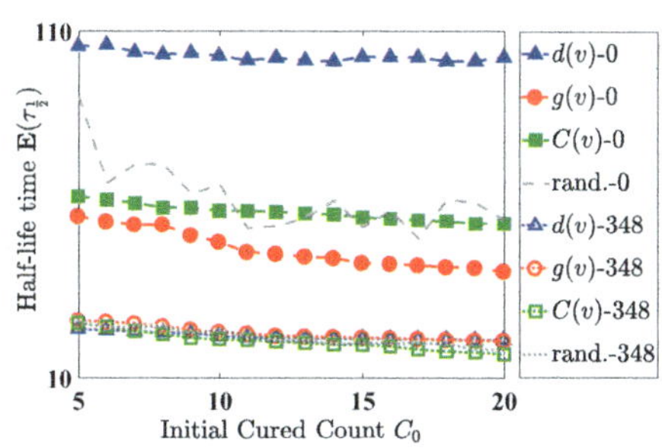

Fig. 2.22 Half-life time $\mathbb{E}(\tau_{\frac{1}{2}})$ on the two networks

which can be clearly observed in Fig. 2.19. This implies the betweenness centrality-based approach will disconnect `dataset0` more easily, while leaving `dataset348` connected during the initial distribution.

Figures 2.21 and 2.22 show the mean extinction time and half-life time, each over 1000 simulation runs. The propagation parameters of the SIC epidemic are:

infection rate $\beta = 0.01$, curing rate $\gamma = 0.01$, and initial infection count $I_0 = |\mathcal{I}(0)| \simeq \frac{1}{2}|\mathcal{V}|$, i.e., 150 for `dataset0` and 110 for `dataset348`. Numerical results in Figs. 2.21 and 2.22 echo with the prediction we made from Fig. 2.20, in which performance of the degree-based approach is poor, and the best approach is either betweenness centrality-based or clustering coefficient-based. In terms of the extinction time, the degree-based (tagged '$d(v)$', blue lines with triangle markers) approach is even worse than random distribution (grey lines), because high degree does not necessarily imply high importance. Due to the clustering effect of human social interactions, high-degree vertices are often located in a densely connected core, where the removal of such a vertex can be compensated by its (usually also high-degree) neighbors. However, the zig-zag pattern of the grey lines, i.e. the random distribution, indicates the instability of this approach. What's more, considering the clustering coefficient-based approach only requires knowledge of two-hop neighbors, rather than global knowledge in the betweenness centrality-based approach, it is an optimal choice, especially when the network is denser. As for the half-life time, which indicates the effectiveness in alleviating heavy infection conditions, simulation suggests similar conclusion as the extinction time, except that degree-based distribution performs better when C_0 is small in `dataset348`, where the network is more clustered.

2.6 Dynamics in Motion: Estimating the Number of Information Adopters at Time *t*

In previous sections, we study the lifetime of the undesired information (virus), which is a 'coarse' way to describe the propagation process through two pivots in time at the system level. To answer the third research question, that is, *how the number of information adopters changes over time*, the dynamic process needs to be analyzed in finer grains. So in this section, we discuss the step-by-step evolution of a SIC epidemic process in a discrete time system with time-steps of length Δt.

We set Δt to be sufficiently short, such that within a time step, e.g., from the $(t-1)$th-step to tth-step, the state of a vertex $i \in \mathcal{V}$, $X_i(t)$ depends solely on the last state of itself $X_i(t-1)$, and that of its neighbors $\{X_j(t-1) | j \in \mathcal{N}(i)\}$. This behavior resembles the *Local Markov property* of Markov Random Fields (MRF), where the (temporal) Markov property is correlated with spatial dependence. To overcome the challenge caused by this correlation, we separate the spatial dependence and temporal dependence through a similar method in [10], but addressing both the virus *infection* and antidote *curing* events at the same time. In this way, we are able to derive a time-recursive expression for the infection probability $\mathbb{P}(X_i(t) = 1)$, by which the expected number of information adopters $\mathbb{E}(I(t))$ and $\mathbb{E}(C(t))$ can be inferred.

Remark 2.1 As we shift our focus from system-level behavior to step-by-step evolution, it is no longer reasonable to employ a unified infection rate β and curing rate γ[12] for all edges. In other words, to obtain a finer time resolution for the dynamic process, we also need a more detailed network model, which captures the difference between two edges. For example, the probability $\beta_{i,j}$ that a virus propagates through edge $e(i, j)$ in a time step may differ from $\beta_{u,v}$ that a virus propagates through edge $e(u, v)$. Considering this change of resolution, throughout this section, we write $\boldsymbol{\beta} := \{\beta_{i,j}\}_{i,j\in\mathcal{V}}$ and $\boldsymbol{\gamma} := \{\gamma_{i,j}\}_{i,j\in\mathcal{V}}$, as the matrix of infection probabilities and curing probabilities, respectively. In this sense, the network $\mathcal{G}(\mathcal{V}, \mathcal{E})$ is heterogeneous, and can be characterized by matrices $\boldsymbol{\beta}$ and $\boldsymbol{\gamma}$.

2.6.1 Temporal Dependence

First, we discuss the state dependence and evolution in the time domain. Let $\mathbb{P}_{uv}(i, t) = \mathbb{P}(X_i(t+1) = v|X_i(t) = u)$ denote the transition probability of vertex i from state u to v during one time step at time t. With states of i's neighbors at time t, the probability that a susceptible vertex i remains susceptible in the next time step $t+1$, or $\mathbb{P}(X_i\left[t+1) = 0 \mid X_i(t) = 0,\ X_{\mathcal{N}(i)}(t) = x_{\mathcal{N}(i)}(t)\right]$, can be written as

$$\mathbb{P}_{00}(i, t) = \prod_{j\in\mathcal{N}(i)} (1-\beta_{j,i})^{\frac{1}{2}(x_j^2(t)+x_j(t))} \cdot (1-\gamma_{j,i})^{\frac{1}{2}(x_j^2(t)-x_j(t))}. \tag{2.62}$$

Further, let $I_i(t) = 1-\prod_{j\in\mathcal{N}(i)}(1-\beta_{j,i})^{\frac{1}{2}(x_j^2(t)+x_j(t))}$, $C_i(t) = 1-\prod_{k\in\mathcal{N}(i)}(1-\gamma_{k,i})^{\frac{1}{2}(x_k^2(t)-x_k(t))}$, then the evolution can be described by the following one-step transition probabilities.

$$\mathbb{P}_{00}(i, t) = \sum_{x_{\mathcal{N}(i)}(t)} \mathbb{P}(X_{\mathcal{N}(i)}(t) = x_{\mathcal{N}(i)}(t)|X_i(t) = 0) \cdot (1-I_i(t))(1-C_i(t)), \tag{2.63}$$

$$\mathbb{P}_{01}(i, t) = \sum_{x_{\mathcal{N}(i)}(t)} \mathbb{P}(X_{\mathcal{N}(i)}(t) = x_{\mathcal{N}(i)}(t)|X_i(t) = 0)I_i(t)(1-C_i(t)), \tag{2.64}$$

$$\mathbb{P}_{11}(i, t) = \sum_{x_{\mathcal{N}(i)}(t)} \mathbb{P}(X_{\mathcal{N}(i)}(t) = x_{\mathcal{N}(i)}(t)|X_i(t) = 1)(1-C_i(t)). \tag{2.65}$$

Note that only three equations are needed to describe the dynamics, because $\sum_{v\in\Lambda} \mathbb{P}_{u,v} = 1$, while $\mathbb{P}_{10} = 0$ and $\mathbb{P}_{-1-1} = 1$. Temporal dependences are included in $I_i(t)$ and $C_i(t)$, while the spatial dependence is captured by the joint conditional probability $\mathbb{P}(X_{\mathcal{N}(i)}(t) = x_{\mathcal{N}(i)}(t)|X_i(t) = x_i(t))$, which is also the

[12] Note that unified infection/curing rate also implies unified infection/curing probability during one unit of time step of length Δt, as discussed in Eq. (2.3) of Sect. 2.2.3.2.

reason that calculating the exact marginal probability distribution is expensive. A common assumption is that during one time step, states of different vertices are mutually independent, i.e.,

$$P(X_{\mathcal{N}(i)}(t) = x_{\mathcal{N}(i)}(t)|X_i(t) = x_i(t)) = \prod_{j\in\mathcal{N}(i)} \mathbb{P}(X_j(t) = x_j(t)), \tag{2.66}$$

but this assumption creates a noticeable deviation in probability, as indicated by [10], so we further discuss spatial dependence in the next step.

2.6.2 *Spatial Dependence*

During one time step, virus and antidote can only propagate to vertices that are at most one hop away, so only the states of one's neighbors are relevant in determining the next state of a vertex. In other words, we can write the temporal and spatial dependence as follows:

$$X_u(t+1) \perp X_{\mathcal{V}\setminus\mathcal{N}(u)\cup\{u\}}(t) \mid X_{\mathcal{N}(u)}(t). \tag{2.67}$$

Based on this observation, we assume the states of vertex i's neighbors $\mathcal{N}(i)(t)$ are independent of each other for every time step t, so

$$\mathbb{P}(X_{\mathcal{N}(i)}(t) = x_{\mathcal{N}(i)}(t)|X_i(t) = x_i(t)) = \prod_{j\in\mathcal{N}(i)} \mathbb{P}(X_j(t) = x_j(t)|X_i(t) = x_i(t)). \tag{2.68}$$

Let $N_0^s(i,j,t) = \mathbb{P}(X_j(t) = s|X_i(t) = 0)$, $N_1^s(i,j,t) = \mathbb{P}(X_j(t) = s|X_i(t) = 1)$, $s \in \{0,-1,1\}$. Then Eqs. (2.63), (2.64), and (2.65) can be simplified into

$$\begin{aligned}\mathbb{P}_{00}(i,t) &= \sum_{x_{\mathcal{N}(i)}(t)} \prod_{j\in\mathcal{N}(i)} \big[N_0^{x_j(t)}(i,j,t)\cdot(1-\beta_{j,i})^{\frac{1}{2}(x_j^2(t)+x_j(t))} \\ &\quad \cdot (1-\gamma_{j,i})^{\frac{1}{2}(x_j^2(t)-x_j(t))} N_0^{x_j(t)}(i,j,t)\big] \\ &= \prod_{j\in\mathcal{N}(i)} \big[1 - N_0^1(i,j,t)\beta_{j,i} - \big(1-N_0^0(i,j,t)-N_0^1(i,j,t)\big)\gamma_{j,i}\big]\end{aligned} \tag{2.69}$$

$$\begin{aligned}\mathbb{P}_{01}(i,t) &= \sum_{x_{\mathcal{N}(i)}(t)} \big[1 - \prod_{j\in\mathcal{N}(i)} (1-\beta_{j,i})^{\frac{1}{2}(x_j^2(t)+x_j(t))}\big] \\ &\quad \cdot \prod_{j\in\mathcal{N}(i)} (1-\gamma_{j,i})^{\frac{1}{2}(x_j^2(t)-x_j(t))} N_0^{x_j(t)}(i,j,t)\end{aligned}$$

$$= \prod_{j \in \mathcal{N}(i)} \left[1 - \left(1 - N_0^0(i, j, t) - N_0^1(i, j, t)\right)\gamma_{j,i}\right] - \mathbb{P}_{00}(i, t) \tag{2.70}$$

$$\mathbb{P}_{11}(i, t) = 1 - \sum_{x_{\mathcal{N}(i)}(t)} \left[1 - \prod_{j \in \mathcal{N}(i)} (1 - \gamma_{j,i})^{\frac{1}{2}(x_j^2(t) - x_j(t))}\right] \cdot \prod_{j \in \mathcal{N}(i)} N_1^{x_j(t)}(i, j, t)$$

$$= \prod_{j \in \mathcal{N}(i)} \left[1 - \left(1 - N_1^0(i, j, t) - N_1^1(i, j, t)\right)\gamma_{j,i}\right]. \tag{2.71}$$

With Eqs. (2.69)–(2.71), we have the recursive master equations of the system

$$\mathbb{P}(X_i(t+1) = 0) = \mathbb{P}(X_i(t) = 0) \cdot \mathbb{P}_{00}(i, t), \tag{2.72}$$

$$\mathbb{P}(X_i(t+1) = 1) = \mathbb{P}(X_i(t) = 0) \cdot \mathbb{P}_{01}(t) + \mathbb{P}(X_i(t) = 1) \cdot \mathbb{P}_{11}(i, t). \tag{2.73}$$

Though $N_0^0(i, j, t)$, $N_0^1(i, j, t)$, $N_1^0(i, j, t)$ and $N_1^1(i, j, t)$ are necessary to solve these equations, their closed form solutions are difficult to obtain. However, it is possible to derive recursive expressions with respect to time, as shown below.

$$N_0^0(i, j, t) = \frac{\mathbb{P}(X_j(t) = 0, X_i(t) = 0)}{\mathbb{P}(X_i(t) = 0)}$$

$$= \frac{\mathbb{P}(X_i(t-1) = 0)}{\mathbb{P}(X_i(t) = 0)} \cdot N_0^0(i, j, t-1) \cdot \mathbb{P}_{00}(i, j, t-1) \cdot \mathbb{P}_{00}(j, i, t-1), \tag{2.74}$$

$$N_0^1(i, j, t) = \frac{\mathbb{P}(X_j(t) = 1, X_i(t) = 0)}{\mathbb{P}(X_i(t) = 0)}$$

$$= \frac{\mathbb{P}(X_i(t-1) = 0)}{\mathbb{P}(X_i(t) = 0)} \cdot \mathbb{P}_{00}(i, j, t-1) \cdot \left[N_0^0(i, j, t-1) \cdot \mathbb{P}_{01}(j, i, t-1)\right.$$

$$\left. + (1 - \beta_{j,i}) \cdot N_0^1(i, j, t-1) \cdot \mathbb{P}_{11}(j, i, t-1)\right], \tag{2.75}$$

$$N_1^0(i, j, t) = \frac{\mathbb{P}(X_j(t) = 0, X_i(t) = 1)}{\mathbb{P}(X_i(t) = 1)}$$

$$= \frac{\mathbb{P}_{00}(j, i, t-1)}{\mathbb{P}(X_i(t) = 1)} \cdot \left[\mathbb{P}(X_i(t-1) = 0) \cdot N_0^0(i, j, t-1) \cdot \mathbb{P}_{01}(i, j, t-1)\right.$$

$$\left. + \mathbb{P}(X_i(t-1) = 1) \cdot (1 - \beta_{i,j}) \cdot N_1^0(i, j, t-1) \cdot \mathbb{P}_{11}(i, j, t-1)\right], \tag{2.76}$$

$$N_1^1(i, j, t) = \frac{\mathbb{P}(X_j(t) = 1, X_i(t) = 1)}{\mathbb{P}(X_i(t) = 1)}$$

$$= \frac{\mathbb{P}(X_i(t-1) = 0)}{\mathbb{P}(X_i(t) = 1)} \cdot \left[N_0^0(i, j, t-1) \cdot \mathbb{P}_{01}(i, j, t-1) \cdot \mathbb{P}_{01}(j, i, t-1)\right.$$

$$+ N_0^1(i,j,t-1) \cdot \mathbb{P}_{11}(j,i,t-1) \cdot [\beta_{j,i} + \mathbb{P}_{01}(i,j,t-1) - \beta_{j,i}\mathbb{P}_{01}(i,j,t-1)]\big] + \frac{\mathbb{P}(X_i(t-1)=1)}{\mathbb{P}(X_i(t)=1)} \cdot \mathbb{P}_{11}(j,i,t-1) \cdot \Big[[\beta_{i,j} + \mathbb{P}_{01}(j,i,t-1) - \beta_{i,j}\mathbb{P}_{01}(j,i,t-1)] \cdot N_1^0(i,j,t-1) + N_1^1(i,j,t-1) \cdot \mathbb{P}_{11}(i,j,t-1)\Big], \tag{2.77}$$

where transition probabilities

$$\mathbb{P}_{00}(i,j,t) = \prod_{k \in \mathcal{N}(i) \setminus \{j\}} \left[1 - N_0^1(i,k,t)\beta_{k,i} - \big(1 - N_0^0(i,k,t) - N_0^1(i,k,t)\big)\gamma_{k,i}\right], \tag{2.78}$$

$$\mathbb{P}_{01}(i,j,t) = \prod_{k \in \mathcal{N}(i) \setminus \{j\}} \left[1 - \big(1 - N_0^0(i,k,t) - N_0^1(i,k,t)\big)\gamma_{k,i}\right] - \mathbb{P}_{00}(i,j,t), \tag{2.79}$$

$$\mathbb{P}_{11}(i,j,t) = \prod_{k \in \mathcal{N}(i) \setminus \{j\}} \left[1 - \big(1 - N_1^0(i,k,t) - N_1^1(i,k,t)\big)\gamma_{k,i}\right], \tag{2.80}$$

denotes the probability that without considering vertex $j \in \mathcal{N}(i)$, vertex i remains susceptible, becomes infected, and remains infected during time step t, respectively.

2.6.3 Expected Infection Count $\mathbb{E}(I(t))$ and Cured Count $\mathbb{E}(C(t))$

Note that Eqs. (2.74)–(2.77) are time-recursive, allowing us to estimate the state evolution as long as the initial state of the network is known (equivalent to the case that the initial distribution of state vector $\mathbf{X}(0) = \{X_v(0)\}_{v \in \mathcal{V}}$ is a δ-distribution at time $t = 0$). It is clear that $\mathbb{P}(X_v(0) = 1) = 1\ \forall v \in \mathcal{I}(0)$, $\mathbb{P}(X_u(0) = -1) = 1$ for any $u \in \mathcal{C}(0)$ and $\mathbb{P}(X_w(0) = 0) = 1$ for any $w \in \mathcal{S}(0)$. Therefore, $\mathbb{P}(\mathbf{X}(0) = \mathbf{x}(0)) = 1 = \prod_{v \in \mathcal{V}} \mathbb{P}(X_v(0) = x_v(0))$, where $\mathbf{x}(0) = \{x_v(0)\}_{v \in \mathcal{V}}$ is the state vector of the network at time 0, indicating r.v.'s $\{X_v(0)\}_{v \in \mathcal{V}}$ are mutually independent. Hence $N_r^s(j,0) = \mathbb{P}(X_j(0) = s | X_i(0) = r) = \mathbb{P}(X_j(0) = s)$ is determined for any $r, s \in \{0, 1, -1\}$. Then $\mathbb{P}(X_i(t) = 1)$ can be solved iteratively, and the expected infection count at time t can be calculated by

$$\mathbb{E}(I(t)) = \mathbb{E}[\sum_{i \in \mathcal{V}} \mathbb{1}_{\{1\}}(X_i(t))] = \sum_{i \in \mathcal{V}} \mathbb{P}(X_i(t) = 1), \tag{2.81}$$

$$\mathbb{E}(C(t)) = \sum_{i \in \mathcal{V}} \left[1 - \mathbb{P}(X_i(t) = 0) - \mathbb{P}(X_i(t) = 1)\right]. \tag{2.82}$$

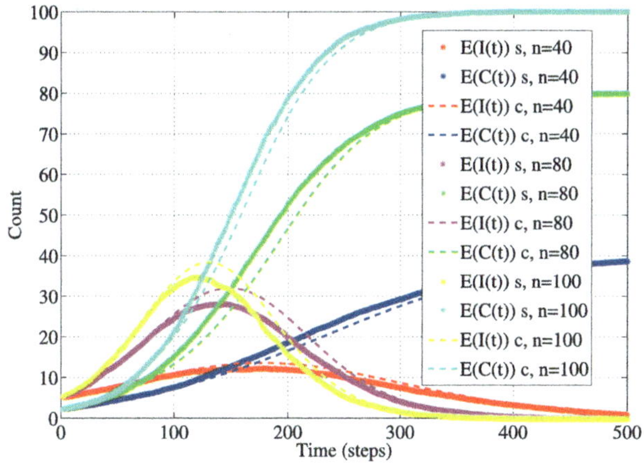

Fig. 2.23 Simulated vs. calculated infection/cured count in networks of size 40, 80 and 100

The manipulations discussed above, whose purposes are to obtain $\mathbb{E}(I(t))$ and $\mathbb{E}(C(t))$ for a given time step t, can be summarized into the following iterative algorithm. Note that $\mathbf{P}_0(t) = \{\mathbb{P}(X_i(t) = 0)\}_{i\in\mathcal{V}}$ and $\mathbf{P}_1(t) = \{\mathbb{P}(X_i(t) = 1)\}_{i\in\mathcal{V}}$ are two $n \times 1$ vectors, while $\mathbf{RS}(t) = \{\mathbb{P}_{00}(i, j, t)\}_{i,j\in\mathcal{V}}$, $\mathbf{NC}(t) = \{\mathbb{P}_{11}(i, j, t)\}_{i,j\in\mathcal{V}}$, $\mathbf{N}_0^0(t) = \{N_0^0(i, j, t)\}_{i,j\in\mathcal{V}}$, $\mathbf{N}_1^0(t) = \{N_0^1(i, j, t)\}_{i,j\in\mathcal{V}}$, $\mathbf{N}_0^1(t) = \{N_1^0(i, j, t)\}_{i,j\in\mathcal{V}}$,$\mathbf{N}_1^1(t) = \{N_1^1(i, j, t)\}_{i,j\in\mathcal{V}}$ are $n \times n$ matrices.

We validate the efficacy of the proposed algorithm in networks with size n ($n \in \{40, 80, 100\}$). In this network, each edge $e(i, j)$ is associated with a random infection/curing probability $\beta_{i,j}$ and $\gamma_{i,j}$. Figure 2.23 shows the simulation (dashed lines with s in the legend) vs. calculation (solid lines composed of dots, with c in the legend) of expected value of infection and cured counts. The infection count $I(t)$ and the curing count $C(t)$ can be viewed as the number of information adopters at time t. As shown by the small gaps between the solid lines and their corresponding dashed lines, $\mathbb{E}(I(t))$ and $\mathbb{E}(C(t))$ can be well captured by the outcome of Algorithm 1.

2.7 Summary

In this chapter, we discuss conflicting information propagation through the lifetime of information in networks. We propose a Susceptible-Infected-Cured propagation model to capture the short-term competitions between the virus (undesired information) and the later-injected antidote (undesired information), both propagating on the same network. We find that the lifetime of the virus can be upper bounded by two topological properties of the network $\mathcal{G}$: the Cheeger constant $\eta(\mathcal{G})$, which measures the level of 'bottleneckness', and vertex eccentricities $\{\epsilon_{\mathcal{G}}(v)\}_v$, which measure the longest distance between vertex v and any other vertices in the network. Particularly,

Algorithm 1 Iterative inference method

Input: $\mathcal{I}(0), \mathcal{C}(0), \boldsymbol{\beta}, \boldsymbol{\gamma}, \mathcal{V}$
Output: $\mathbb{E}(I(t)), \mathbb{E}(C(t))$
1: Initialize: $n = |\mathcal{V}|, I(0) = |\mathcal{I}(0)|, C(0) = n - I(0) = |\mathcal{C}(0)|$
$\mathbf{P}_0(0) = \mathbf{N}_0^0(0) = \mathbf{N}_1^0(0) = \mathbb{1}_{\mathcal{V}\setminus(\mathcal{I}(0)\cup\mathcal{C}(0))}$,
$\mathbf{P}_1(0) = \mathbf{N}_0^1(0) = \mathbf{N}_1^1(0) = \mathbb{1}_{\mathcal{I}(0)}$
2: Calculate $\mathbf{P}_0(1)$ with Eqs. (2.65) and (2.71)
3: Calculate $\mathbf{P}_1(1)$ with Eqs. (2.69), (2.70), and (2.73)
4: Output $\mathbb{E}(I(1)) = sum(\mathbf{P}_1(1))$,
$\mathbb{E}(C(1)) = n - \mathbb{E}(I(1)) - sum(\mathbf{P}_0(1))$
5: Calculate $\mathbf{RS}(0)$ and $\mathbf{NC}(0)$ with Eqs. (2.78) and (2.79)
6: $t = 1$
7: **while** $\mathbb{E}(C(t)) < n$ **do**
Calculate $\mathbf{N}_0^0(t), \mathbf{N}_0^1(t), \mathbf{N}_1^0(t), \mathbf{N}_1^1(t)$ with Eqs. (2.74)–(2.78)
Calculate $\mathbf{P}_0(t+1)$ with Eqs. (2.65) and (2.71)
Calculate $\mathbf{P}_1(t+1)$ with Eqs. (2.69), (2.70), and (2.73)
Output $\mathbb{E}(I(t+1)) = sum(\mathbf{P}_1(t+1))$,
$\mathbb{E}(C(1)) = n - \mathbb{E}(I(t+1)) - sum(\mathbf{P}_0(t+1))$
Calculate $\mathbf{RS}(t)$ and $\mathbf{NC}(t)$ with Eqs. (2.78) and (2.79)
$t = t + 1$
8: **end while**

the $O(\frac{\log n}{\eta(\mathcal{G})})$ bound indicates that the lifetime of the virus does not always decrease with network size n, while the upper bounds with $\{\epsilon_{\mathcal{G}}(v)\}_v$ are less computationally expensive to obtain. As an application of this knowledge, we provide practical antidote distribution strategies with real-world network traces. Finally, we propose an inference algorithm to estimate the number of information adopters, which can foresee the instantaneous evolution of the dynamics, before it fully unfolds. As the driving force of mobile data, information dynamics determines *when* data moves and stops, as well as the volume of data traffic in networks. Knowledge on the lifetime of information provides guidelines on network topology design, data traffic prediction, and control of undesired information propagation.

References

1. Alexander, H.: Boston bomb: Reddit apologises for identifying wrong suspects. http://www.telegraph.co.uk/news/worldnews/northamerica/usa/10012382/ (2013)
2. Aven, T.: Upper (lower) bounds on the mean of the maximum (minimum) of a number of random variables. J. Appl. Probab. **22**(3), 723–728 (1985). https://doi.org/10.2307/3213876
3. Beutel, A., Prakash, B.A., Rosenfeld, R., Faloutsos, C.: Interacting viruses in networks: Can both survive? In: Proceedings of the 18th ACM SIGKDD International Conference on Knowledge Discovery and Data Mining, KDD '12, pp. 426–434. ACM (2012). https://doi.org/10.1145/2339530.2339601. http://doi.acm.org/10.1145/2339530.2339601
4. Bienstock, D.: Optimal control of cascading power grid failures. In: 2011 50th IEEE Conference on Decision and Control and European Control Conference, pp. 2166–2173 (2011). https://doi.org/10.1109/CDC.2011.6160415

5. Borgs, C., Chayes, J., Ganesh, A., Saberi, A.: How to distribute antidote to control epidemics. Random Struct. Algorithms **37**(2), 204–222 (2010). https://doi.org/10.1002/rsa.20315. http://dx.doi.org/10.1002/rsa.20315
6. Brandes, U.: A faster algorithm for betweenness centrality. J. Math. Sociol. **25**, 163–177 (2001)
7. Caccetta, L., Smyth, W.: Graphs of maximum diameter. Discrete Math. **102**(2), 121–141 (1992). https://doi.org/10.1016/0012-365X(92)90047-J. http://www.sciencedirect.com/science/article/pii/0012365X9290047J
8. Chakrabarti, D., Wang, Y., Wang, C., Leskovec, J., Faloutsos, C.: Epidemic thresholds in real networks. ACM Trans. Inf. Syst. Secur. **10**(4), 1:1–1:26 (2008). https://doi.org/10.1145/1284680.1284681. http://doi.acm.org/10.1145/1284680.1284681
9. Chen, P.Y., Chen, K.C.: Optimal control of epidemic information dissemination in mobile ad hoc networks. In: Global Telecommunications Conference (GLOBECOM 2011), 2011 IEEE, pp. 1–5 (2011). https://doi.org/10.1109/GLOCOM.2011.6134328
10. Chen, Z., Ji, C.: Spatial-temporal modeling of malware propagation in networks. IEEE Trans. Neural Netw. **16**(5), 1291–1303 (2005). https://doi.org/10.1109/TNN.2005.853425
11. Chung, F.R.: Laplacians of graphs and Cheeger's inequalities. In: Combinatorics, Paul Erdos is Eighty, vol. 2, no. 157–172, pp. 13–2 (1996)
12. Cisco: Cisco Visual networking index: global mobile data traffic forecast update, 2017–2022. Tech. rep., 2019
13. Dadlani, A., Kumar, M.S., Maddi, M.G., Kim, K.: Mean-field dynamics of inter-switching memes competing over multiplex social networks. IEEE Commun. Lett. **21**(5), 967–970 (2017). https://doi.org/10.1109/LCOMM.2017.2651815
14. Dasarathy, G.: A simple probability trick for bounding the expected maximum of n random variables. https://www.ece.rice.edu/~gd14/files/maxGaussians.pdf (2011)
15. De Marneffe, M.C., Rafferty, A.N., Manning, C.D.: Finding contradictions in text. In: ACL, vol. 8, pp. 1039–1047 (2008)
16. Ganesh, A., Massoulie, L., Towsley, D.: The effect of network topology on the spread of epidemics. In: INFOCOM 2005. 24th Annual Joint Conference of the IEEE Computer and Communications Societies. Proceedings IEEE, vol. 2, pp. 1455–1466 (2005). https://doi.org/10.1109/INFCOM.2005.1498374
17. Garnaeva, M., Chebyshev, V., Makrushin, D., Unuchek, R., Ivanov, A.: Kaspersky security bulletin 2014. Overall statistics for 2014. http://securelist.com/analysis/kaspersky-security-bulletin/68010/kaspersky-security-bulletin-2014-overall-statistics-for-2014/ (2014)
18. Hoory, S., Linial, N., Wigderson, A.: Expander graphs and their applications. Bull. Am. Math. Soc. **43**(4), 439–561 (2006)
19. Khouzani, M., Sarkar, S., Altman, E.: Optimal control of epidemic evolution. In: INFOCOM, 2011 Proceedings IEEE, pp. 1683–1691 (2011). https://doi.org/10.1109/INFCOM.2011.5934963
20. Kim, M., Candan, K.S.: SBV-cut: Vertex-cut based graph partitioning using structural balance vertices. Data Knowl. Eng. **72**, 285–303 (2012). https://doi.org/10.1016/j.datak.2011.11.004. http://www.sciencedirect.com/science/article/pii/S0169023X11001480
21. Krishnasamy, S., Banerjee, S., Shakkottai, S.: The behavior of epidemics under bounded susceptibility. In: The 2014 ACM International Conference on Measurement and Modeling of Computer Systems, SIGMETRICS '14, pp. 263–275. ACM, New York (2014). https://doi.org/10.1145/2591971.2591977. http://doi.acm.org/10.1145/2591971.2591977
22. Lee, D.: Boston bombing: How internet detectives got it very wrong. http://www.bbc.com/news/technology-22214511 (2013)
23. Lelarge, M.: Efficient control of epidemics over random networks. In: Proceedings of the Eleventh International Joint Conference on Measurement and Modeling of Computer Systems, SIGMETRICS '09, pp. 1–12. ACM, New York (2009). https://doi.org/10.1145/1555349.1555351. http://doi.acm.org/10.1145/1555349.1555351
24. Leskovec, J., Backstrom, L., Kleinberg, J.: Meme-tracking and the dynamics of the news cycle. In: Proceedings of the 15th ACM SIGKDD International Conference on Knowledge Discovery

and Data Mining, KDD '09, pp. 497–506. ACM, New York (2009). https://doi.org/10.1145/1557019.1557077. http://doi.acm.org/10.1145/1557019.1557077
25. Leskovec, J., Mcauley, J.J.: Learning to discover social circles in ego networks. In: Bartlett, P., Pereira, F., Burges, C., Bottou, L., Weinberger, K. (eds.) Advances in Neural Information Processing Systems, vol. 25, pp. 548–556 (2012). http://books.nips.cc/papers/files/nips25/NIPS2012_0272.pdf
26. Lin, Y., Lui, J., Jung, K., Lim, S.: Modeling multi-state diffusion process in complex networks: Theory and applications. In: Signal-Image Technology Internet-Based Systems (SITIS), 2013 International Conference on, pp. 506–513 (2013). https://doi.org/10.1109/SITIS.2013.86
27. Louis, A.: The complexity of expansion problems. Ph.D. thesis, Georgia Institute of Technology, 2014
28. Miller, M., Siran, J.: Moore graphs and beyond: A survey of the degree/diameter problem. Electron. J. Combin. (Dyn. Surv.), 1–92 (2013). http://oro.open.ac.uk/41753/
29. Milliken, J., Selis, V., Marshall, A.: Detection and analysis of the chameleon WiFi access point virus. EURASIP J. Inf. Secur. **2013**(1), 2 (2013). https://doi.org/10.1186/1687-417X-2013-2. http://jis.eurasipjournals.com/content/2013/1/2
30. Newman, M.E.J.: Threshold effects for two pathogens spreading on a network. Phys. Rev. Lett. **95**(10), 108701 (2005). https://doi.org/10.1103/PhysRevLett.95.108701
31. Nowzari, C., Preciado, V.M., Pappas, G.J.: Analysis and control of epidemics: A survey of spreading processes on complex networks. IEEE Control Syst. Mag. **36**(1), 26–46 (2016). https://doi.org/10.1109/MCS.2015.2495000
32. Pastor-Satorras, R., Vespignani, A.: Epidemic spreading in scale-free networks. Phys. Rev. Lett. **86**(14), 3200 (2001)
33. Prakash, B., Chakrabarti, D., Valler, N., Faloutsos, M., Faloutsos, C.: Threshold conditions for arbitrary cascade models on arbitrary networks. Knowl. Inf. Syst. **33**(3), 549–575 (2012). https://doi.org/10.1007/s10115-012-0520-y
34. Prakash, B.A., Beutel, A., Rosenfeld, R., Faloutsos, C.: Winner takes all: Competing viruses or ideas on fair-play networks. In: Proceedings of the 21st International Conference on World Wide Web, WWW '12, pp. 1037–1046. ACM, New York (2012). https://doi.org/10.1145/2187836.2187975. http://doi.acm.org/10.1145/2187836.2187975
35. Preciado, V., Zargham, M., Sun, D.: A convex framework to control spreading processes in directed networks. In: Information Sciences and Systems (CISS), 2014 48th Annual Conference on, pp. 1–6 (2014). https://doi.org/10.1109/CISS.2014.6814167
36. Qi, F., Guo, B.N.: Sharp bounds for harmonic numbers. Preprint. arXiv:1002.3856 (2010)
37. Takes, F.W., Kosters, W.A.: Computing the eccentricity distribution of large graphs. Algorithms **6**(1), 100–118 (2013). https://doi.org/10.3390/a6010100. https://www.mdpi.com/1999-4893/6/1/100
38. VNI, C.: Cisco visual networking index: forecast and trends, 2017–2022 White Paper. Tech. rep., Cisco and/or its affiliates, 2019
39. Wang, J.W., Rong, L.L.: Cascade-based attack vulnerability on the us power grid. Saf. Sci. **47**(10), 1332–1336 (2009)
40. Wiener, J., Bronson, N.: Facebook's top open data problems. https://research.fb.com/facebook-s-top-open-data-problems/ (2014)
41. Yang, J., Leskovec, J.: Patterns of temporal variation in online media. In: Proceedings of the Fourth ACM International Conference on Web Search and Data Mining, WSDM '11, pp. 177–186. ACM, New York (2011). https://doi.org/10.1145/1935826.1935863. http://doi.acm.org/10.1145/1935826.1935863
42. Zhang, H., Zhang, Z., Dai, H.: Gossip-based information spreading in mobile networks. IEEE Trans. Wirel. Commun. **12**(11), 5918–5928 (2013). https://doi.org/10.1109/TWC.2013.100113.130619

Chapter 3
Coverage Dynamics: Modeling and Analysis of Data Coverage in Heterogeneous Edge Networks

Our study in the previous chapter revealed the *lifetime* of mobile data in large networks, to address the *when* question. In this chapter, we shift gear to another important aspect of mobile data, that is, the *whereabouts* of data in the space domain. Specifically, we study *where* data are accessible, i.e., *data coverage* during a data dissemination process. This question is of particular importance to location-based service (LBS), in which data are tagged with geographical positions, and disseminated though mobile devices with diverse capabilities in edge networks. However, existing research on data dissemination either do not carry the notation of 'where', or assume participants of the dissemination process to be homogeneous in communication, mobility, and other aspects. To overcome these challenges induced by the introduction of spatial domain and device heterogeneity, we propose a data coverage model that captures the attributes and actions of individual participants, and quantify the data coverage with a *data strength* metric, by which the dynamic change of data coverage can be described as a numeric signal on a *graph* that embeds location information. Analyzing this graph signal over entity mobility and data forwarding, we find that, the impact of entity mobility is high, especially when entities are less active in forwarding, or moving in low speed. Moreover, such impact acts in the form of a user mobility matrix, which indicates the underlying graph topology of the defined signal of data strength.

3.1 Motivation and Related Work

Recent advances in data-driven applications are re-enforcing the indispensable role of rapidly growing *mobile data* [3] in a variety of scenarios, ranging from marketing [22] to manufacturing [20], and scientific research [4]. In the era of mobile data, generation, dissemination and consumption of data are much easier than ever, owing to the proliferation of increasingly powerful mobile devices, such as smart phones

J. Wang et al., *Encountering Mobile Data Dynamics in Heterogeneous Wireless Networks*, https://doi.org/10.1007/978-3-031-62906-8_3

[25], as well as emerging networking paradigms, such as Internet-of-Things (IoT) and fog computing [2, 12]. Consequently, data dissemination processes, which move data from their generator(s) to their consumer(s), as if data are a special type of commodities, are quickly migrating to the *edge* networks, which are composed of both user-operated wireless devices that can move, such as smart phones, wearable devices, and smart vehicles; and stationary network elements that provide wireless access to devices as a part of the infrastructure, such as cellular BS, LTE eNodeB, and WiFi AP.

3.1.1 Motivation

Such data dissemination services are imperative to a lot of applications. For instance, in mobile advertisement (ad) applications [13, 21], business-owners can recruit smart phones, nearby vehicles and AP's to distribute ads, such that an ad penetration is maintained over populous regions. Another example is traffic information collection/distribution in intelligent transportation systems (ITS) [1, 23], which heavily relies on communication among smart vehicles with on-board unit (OBU), IoT sensors, RSU, and cellular eNodeB. Despite their differences, data dissemination processes in these applications share some common grounds:

1. *Heterogeneous network*. Data are generated and disseminated by mobile phones, smart vehicles, AP's, and BS's, which we refer to as *entities*, with diverse communication schemes, mobility patterns, and forwarding preferences. For example, the communication range of a BS (100 m–10 Km) is much larger than that of an AP (20 m–1 Km), and a pedestrian, who streams video over his/her smart phone, moves much slower than a vehicle on the highway, whose registration information is collected and routed by the electric toll collecting (ETC) sensors. As a result, entities may intermittently connect, disconnect, and re-connect as they move, so data to be disseminated traverse a *heterogeneous* network via both wired (e.g., the X2 connection between LTE eNodeB's and wireless (e.g., the D2D connection between smart phones) links, before they are picked up by consumers.
2. *Mobile data*. Ever since its injection into the edge network, a data block (which we refer to as *datum* throughout this chapter) and its copies are constantly *moving*, due to user/entity movements such as smartphone users roaming in a park, data forwarding actions such as ad broadcasting along a road, and entities' decisions to join/withdraw from the data dissemination process.
3. *Location-based service (LBS)*. The dissemination service is location-centric, because *where* data are determines *who* can immediately access the data. This is especially evident to applications that deal with location-tagged data, e.g., traffic information collected and exchanged among smart vehicles, sensory data collected and actuation commands distributed in IoT, etc.

A key question arises from the dissemination process of data: *where are the accessible data*? Answer to the 'where' question is meaningful to multiple stakeholders of a data dissemination service, and can effectively guide a number of design issues, such as placement of network elements, allocation of storage/processing resources, and evaluation of network performances. From the perspective of data owners, e.g., companies who purchase mobile ad service from AdMob, it is their legitimate rights to know where their data can be accessed, and preferably accessed by whom. To data service providers, e.g.,content delivery network (CDN) like Cloudflare, whereabouts of data should be disclosed to content owners for service transparency, and as input to servers deployment planning. In regard to the network operators, e.g., cellular carriers like AT&T, changing volume of mobile data translate to traffic engineering, and management of network resources, which should be considered in network planning, charging, and OA&M (operation, administration, and maintenance). Therefore, it is necessary for all parties, to have a quantitative measure on the *whereabouts* of data in dissemination, and preferably a mechanism to evaluate data accessibility over a geographical region of interest.

3.1.2 Related Work

Data dissemination process in wireless networks has been extensive studied from the following aspects, such as data dissemination time [24], throughput [6], and network topology [13]. Among these, Grossglaussor and Tse proved movements of entities (under speed-constrained mobility) can increase the system capacity (in terms of pairwise throughput) for ad hoc networks [6] relying uni-casts. Under a move-and-gossip dissemination model, Zhang et al. defined a mobile conductance metric, to derive the spreading time, by when all of the entities have received the data [24]. Li et al. studied the speed of data propagation with geo-cast [7] in a group of mobile devices following the same propagation mode and mobility pattern. Lu et al. showed there exists a critical condition characterized by participating probability, moving speed and transmission range, above which the number of data-carrying entities will scale quadratically with time [10]. In the aforementioned literature, participants of the data dissemination process are assumed to be homogeneous in aspects including communication capacity, mobility pattern, and data forwarding preferences. This is not the case in the edge network, where participants can be drastically different in these aspects. Taking heterogeneity in communication protocols into consideration, hierarchical topology of mobiles in a cellular and WiFi overlay network is studied in [8]. From the perspective of data forwarding influenced by social features, Qin et al. analyzed the dynamic sociality of vehicles, based on which they proposed a mobile advertisement dissemination scheme that can achieve a shorter distributing time [14]. Despite it significant impact, the 'where' question remains open, as the aforementioned literature do not carry the notion of data's whereabouts.

3.1.3 *Our Approach and Contributions*

Motivated by the lack of study on the whereabouts of data, this chapter discusses *data coverage*, that is, the time-varying locations where a data block of interest can be accessed during its dissemination process. However, answer to the 'where' question are hindered by two main challenges arising from the edge network scenario: First, from the unique setting of data dissemination processes, how to account for the heterogeneity among entities, including both mobile devices, and stationary access network elements that are different in many aspects? Second, how to quantitatively describe the whereabouts of a data block (and its copies), considering that they are constantly moving over-the-air due to forwarding, being downloaded and/or deleted by users, and piggybacking on moving entities?

To overcome the first challenge of heterogeneity, we propose a generic entity model that captures the attributes and actions of participants in a dissemination process, and formally define a measurable scope of *data coverage*, i.e., the geographical region in which the data block can be accessed to answer the 'where' question. Then to address the second challenge of quantification, we define a new metric, *data-strength*, through which the *coverage dynamics*, i.e., time-varying data coverage, can be quantitatively described by a signal on a *graph* that embeds geographical location information. The graph signal formulation allows us to study the evolution of the dynamic process of data movement and coverage. Particularly, we analyze the impact of user mobility and data forwarding through a *snapshot* of the coverage dynamics, i.e., the difference graph signal, outlining the possibility of its prediction. In summary, this chapter makes the following contributions toward understanding mobile data dissemination in the edge networks:

We formally define the data movement and coverage problem, propose a new metric, data-strength, to characterize the dissemination process, and formulate data coverage as a tractable time series of numeric graph signals.

We observe bounded impact from user mobility and data forwarding on the change of data-strength, as the maximum moving speed or participating ratio increases. Particularly, the impact of user mobility is high when entities are not active in moving or data forwarding, and this impact takes effect in the form of a mobility matrix, which describes the underlying graph of the data strength signal. It indicates that data coverage in the future can be predicted based on the current data strength, and the long-term mobility patterns.

3.2 Problem Formulation

As discussed in the previous section, existing research on data dissemination focus on 'how many' have received the data, instead of 'where' data can be accessed. To address this issue, we introduce the *data coverage* model to formulate the intuitive 'where' question into specific and rigorous research questions. We first specify the scope of the 'where' question, establish the entity model to characterize dissemination participants of various kinds, and formally define coverage dynamics.

3.2.1 Scope of the 'Where' Problem

The time-varying whereabouts of data due to entity movements and data forwarding are relevant to three domains: time, space, and data.

1. *Time*: The system runs in continuous time $[0, \infty)$, and is observed every Δ time. The 0-th observation taken at time instant 0 is called the initial state of the system. For integer $t \in \mathcal{T} = \{1, 2, \cdots\}$, we differentiate the following:
   - *Time (instant) t* refers to the time *instant* $t \cdot \Delta \in [0, \infty)$, when we take the t-th observation of the dissemination process.
   - *Time step t* refers to the time *interval* $[(t-1)\Delta, t\Delta) \subset [0, \infty)$, during which actions and interactions of entities take place in the system.

   With this clarification, we consider intervals of unit length ($\Delta = 1$), and write $t \in \mathcal{T}$ throughout this chapter, when no confusion is raised.
2. *Space*: Consider a planar region $A \subset \mathbb{R}^2$, where mobile ads are disseminated through smart phones. Let $d_A(\cdot, \cdot)$ denote the Euclidean distance in A.
3. *Data*: Let D denote a set of distinct data blocks. We consider the dissemination process of one data block $\delta \in D$, later referred to as a *datum*, which has been injected into the network before time instant $t = 0$. The dissemination process of datum δ refers to the replication, forwarding, and deletion of δ that take place in region A. Exact copies of datum δ (as a result of replication) are treated as the same data block as the original datum δ during the dissemination process, but any distorted, or altered copies of δ are viewed as different datum than δ, and is hence not considered in the dissemination process.

Participants of a data dissemination process include mobile wireless devices, e.g., smart phones, IoT sensors, and OBU's, as well as infrastructure nodes in wireless access networks, e.g., AP's, eNodeB's, and RSU's. They compose a group of *entities* E, each of which can receive, carry, distribute, and delete datum δ at any time.

3.2.2 Entity Model

Each entity $e \in E$ is characterized by a list of attributes and a set of actions that can change entity e's own dynamic attributes (and hence the whereabouts of datum δ).

3.2.2.1 Attributes of Entity *e*

Attributes of an entity describe its (static) capability and (dynamic) statuses.

1. *Transmission range* $r_e \in \mathbb{R}^+$ is defined as the maximum distance that a data block $\delta \in D$ can be transmitted successfully over wireless links by entity e in one time step. It is a static attribute that does not change over time.

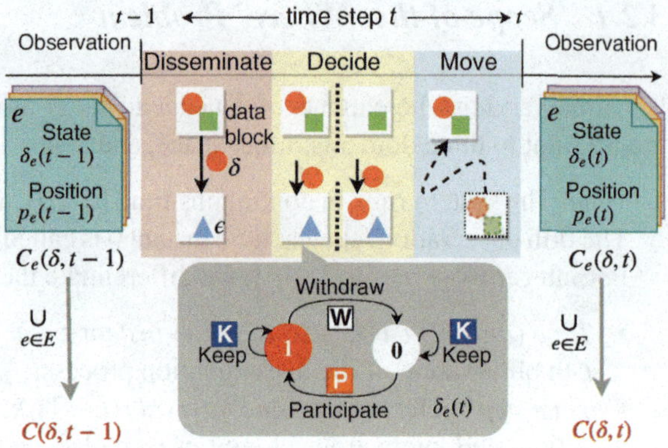

Fig. 3.1 Data coverage model (Note that time is not to scale in this diagram)

2. *Position* $p_e(t) \in \mathbb{R}^2$ records the exact location of entity e, e.g., GPS coordinates, observed at time t. It is a dynamic status. Note that it is possible for an entity e to move out of (and back into) region A at some time instant.
3. *State* $\delta_e(t) \in \{0, 1\}$ is an indicator of whether entity e has a copy of datum δ at time instant t, as well as a determinant of entity e's actions in the next ($t + 1$-th) time step. Entity e will participate in the dissemination process, i.e., promises to disseminate copies of δ, during time step $t + 1$, if and only if its state $\delta_e(t) = 1$. It is a dynamic status of entity e.

The dynamic statuses of entity e, that is, position $p_e(t)$ and states $\delta_e(t)$ observed at time instant t, are changed by actions e takes in past time step t (as time instant t marks the end of time step t, as shown in Fig. 3.1). Moreover, the current status of e also partially determines the actions e will take during time step $t + 1$.

3.2.2.2 Actions

During every time step $t \geq 1$, each entity takes three actions sequentially: one disseminate action, one decide action, and one move action, as illustrated in Fig. 3.1.

1. *Disseminate*: At the beginning of time step t, given its previous state $\delta_e(t - 1)$, entity e will take exactly one of the following two actions:

   - *Forward* datum δ to all possible recipients, if entity e decided to participate (in the previous time step), i.e., $\delta_e(t - 1) = 1$. Data recipients include any entity $x \in E$ within its wireless communication range, that is, $\{x \in E \mid d_A(p_e(t), p_x(t)) \leq \min\{r_c^e, r_c^x\}\}$, and (stationary) entity y that is directly connected with (stationary) entity e via wired links;
   - *Listen* to broadcasts of others, if $\delta_e(t - 1) = 0$.

2. *Decide*: Then entity e sets its state $\delta_e(t)$ to indicate whether e will:
   - *Keep* its current state during time step $t+1$, such that $\delta_e(t) = \delta_e(t-1)$, as shown by the blue boxes with letter K in Fig. 3.1;
   - *Withdraw* from the dissemination process, by deleting the local copy of δ and setting $\delta_e(t) = 0$, given that it was participating in the previous time step, i.e., $\delta_e(t-1) = 1$, as shown by the white box with letter W in Fig. 3.1;
   - *Participate* in the dissemination process of datum δ, by setting state $\delta_e(t) = 1$, as shown by the orange box with letter P in Fig. 3.1, given that (1) it was not participating in the last time step, i.e., $\delta_e(t-1) = 0$, and (2) it has received a copy of δ during the dissemination stage in time step t.
3. *Move*: Finally entity e moves from its last position $p_e(t-1)$ to a new position $p_e(t) \in \mathbb{R}^2$. For stationary entities, such as an eNodeB, we write $p_e(t) = p_e(t-1), \forall\, t \in \mathcal{T}$, so that all entities can be described by the same model.

3.2.3 Data Coverage and Coverage Dynamics

Due to the movements, actions, and interactions of entities in E, the region where a datum can be accessed, i.e., its data coverage, changes over time.

Definition 3.1 (Data) Coverage $C(\delta, t)$ **of datum** δ at time t is defined as a subregion of region A, where a copy of δ can be obtained in time step $t+1$, that is,

$$C(\delta, t) := \{a \in A \mid \exists\, e \in E,\ s.t.\ a \in C_e(\delta, t)\}, \tag{3.1}$$

where $C_e(\delta, t)$ denotes the individual coverage offered by entity e, i.e., locations where δ can be obtained from e, that is,

$$C_e(\delta, t) := \begin{cases} \{a \in A \mid d_A(a, p_e(t)) \leq r_e\}, & \text{if } \delta_e(t) = 1, \\ \phi, & \text{otherwise.} \end{cases} \tag{3.2}$$

An Example Figure 3.2, 3.3, 3.4, and 3.5 illustrate the time-varying data coverage (red shaded area) in a system of eight entities, two (access point (AP) e_1 and eNodeB e_6) of which are stationary entities connected by wired links. During this (short) dissemination process, the system is observed at time instant $t = 0$ and $t = 1$, as

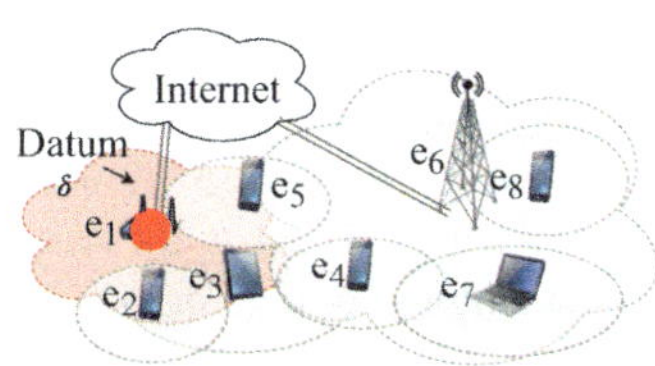

Fig. 3.2 Initial state of the coverage dynamics observed at $t = 0$

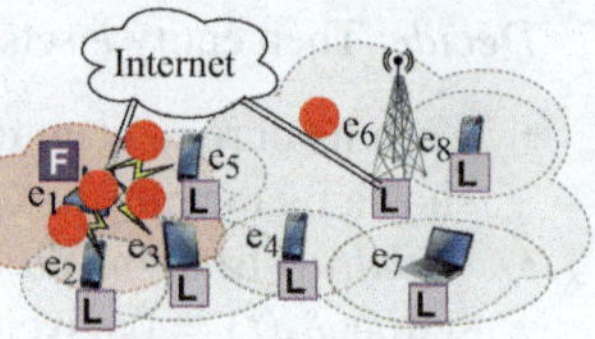

Fig. 3.3 Data dissemination stage during time step $t = 1$

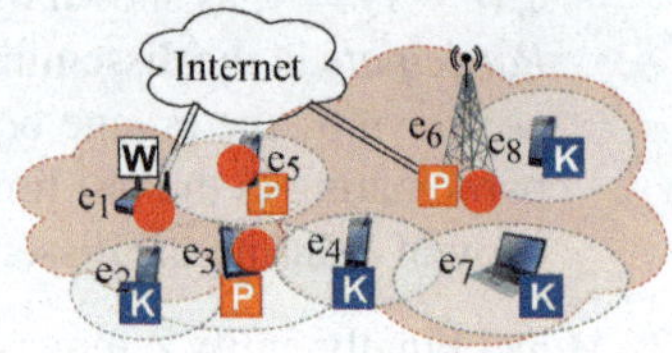

Fig. 3.4 Individual decision stage during time step $t = 1$

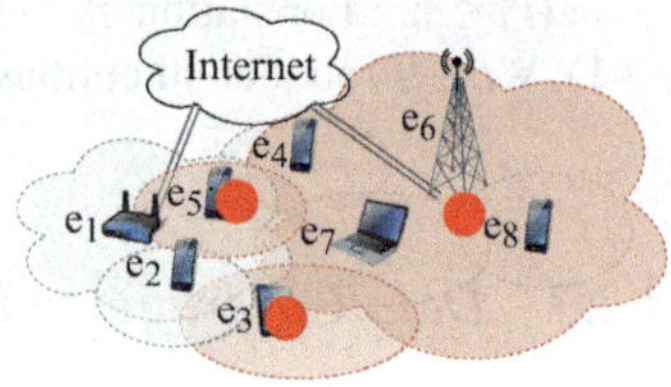

Fig. 3.5 Next state: observation at time $t = 1$

shown in Figs. 3.2 and 3.5. During time step 1, the following individual actions drive the coverage dynamics, i.e., changes the data coverage:

(a) By time instant $t = 0$, datum δ has been injected into the system through entity e_1, so $\delta_{e_1}(0) = 1$. Data coverage $C(\delta, 0) = C_{e_1}(\delta, 0)$, and $C_{e_i}(\delta, 0) = \phi$ for $i \neq 1$.
(b) At the beginning of time step 1, that is, the dissemination stage, entity e_1 forwards datum δ to potential recipients $\{e_2, e_3, e_5, e_6\}$, while all other entities listen, as shown by the light purple boxes with letter L in Fig. 3.3. Note that a copy of δ is passed to entity e_6 via the wired link as well.
(c) Then in the decision stage in Fig. 3.4, entity e_3, e_5, e_6 decide to participate ($\delta_{e_3}(1) = \delta_{e_5}(1) = \delta_{e_6}(1) = 1$); entity e_2, e_4, e_7 and e_8 decide to keep their current states ($\delta_{e_2}(1) = \delta_{e_4}(1) = \delta_{e_7}(1) = \delta_{e_8}(1) = 0$); while entity e_1 decides to withdraw from the dissemination ($\delta_{e_1}(1) = 0$).
(d) At time instant $t = 1$, the system is observed again (after each mobile entity moves) for data coverage $C(\delta, 1)$, which equals to $\cup_{i=3,5,6} C_{e_i}(\delta, 1)$.

In this way, the coverage of datum δ changes over time, snapshots of which compose the *coverage dynamics* $\{C(\delta, t)\}_{t \in \mathcal{T}} \subset A^{\mathcal{T}}$ that describes the *whereabouts* of datum δ during its dissemination process. Based on this data coverage model, we restate the 'where' question into two sub-questions with respect to $\{C(\delta, t)\}_{t \in \mathcal{T}}$:

1. *Quantification*: How to describe and calculate data coverage $C(\delta, t)$ at time t?
2. *Evolution*: What is the impact of user mobility and data forwarding actions on the change of data coverage over a time step, e.g., from $C(\delta, t)$ to $C(\delta, t + 1)$?

The first question calls for a quantification of coverage dynamics, which is a premise to address the other two problems. Therefore in the next section, we introduce our analysis framework that formulates the dynamics into a structured set of numeric random processes, before addressing the evolution problem.

3.3 Representing Coverage Dynamics with Graph Signals

By definition, data coverage of a datum is the union of individual coverages. Though intuitive, this entity-centric definition with the union operator does not naturally provide any quantitative measure on coverage. In other words, we can not calculate, or compare data coverage instances. For example, we can not compare coverage of datum δ at different time instants, $C(\delta, t)$ and $C(\delta, t+1)$, or compare the coverage of different data blocks at the same instant, e.g., $C(\epsilon, t)$ of ϵ and $C(\delta, t)$ of δ.

To address this *quantification* issue, we make coverage dynamics tractable by a numeric measure, and then discuss entity-level state transitions, and system-level governing equations, to form the analysis framework of data coverage dynamics.

3.3.1 A Location-Centric Measure: Data-Strength

The cornerstone of this framework is a metric to represent data coverage $C(\delta, t)$ numerically, by which the coverage (a set of infinite location points) of an irregular shape, e.g., $C(\delta, 1)$ in Fig. 3.5 can be quantitatively compared with its predecessor $C(\delta, 0)$ in Fig. 3.2. Ideally, this metric should reflect the *strength* of datum δ's coverage at time t. For example in Fig. 3.6 (left), darkness of the shade indicates the number of datum copies that may be obtained at different locations: the darker the shade, the more probable that a copy of datum can be obtained.

To do so, we change from an entity-centric perspective, to a location-centric perspective. First, partition region A into a grid of N non-overlapping sub-regions (later referred to as *cells*) $\{A_1, A_2, \cdots, A_N\}$, as shown in Fig. 3.6 (middle). Then for each cell A_n, measure the **data-strength**, defined as

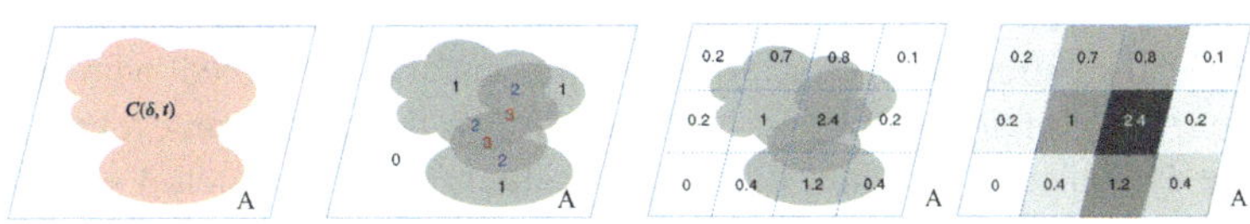

Fig. 3.6 Data-strength $\mathbf{s}(\delta, t)$ describes data coverage $C(\delta, t)$ at the space granularity of cells: Data coverage $C(\delta, t)$ of irregular shape can be quantified with the strength in terms of number of datum copies. Then partitioning region A into a grid of $N = 16$ square cells allows us to describe the spatial coverage with data-strength $\mathbf{s}(\delta, t) \in \mathbb{R}_0^N$, which is a vector of N real numbers

$$s^n(\delta, t) := \sum_{e \in E} \frac{|A_n \cap C_e(\delta, t)|}{|A_n|}, \tag{3.3}$$

where $|\cdot|$ denotes the area of a region.

The the assembled vector of data strength, $\mathbf{s}(\delta, t) = [s^1, s^2, \cdots, s^N](\delta, t)$, can fully describe data coverage at time t, in the sense that:

1. The location information (*where*) is encoded in the superscripts of vector $\mathbf{s}(\delta, t)$, such that changes in space of A are transformed into variation of numeric values;
2. Given vector $\mathbf{s}(\delta, t)$, data coverage $C(\delta, t)$ can be approximated by the union of cells in set $\{A_n \mid s^n(\delta, t) > \alpha\}$, where $\alpha \geq 0$ is a predetermined threshold;
3. The data-strength metric measures the expected number of copies of datum δ an entity can gather if it is located in a cell. So data-strength of a cell can be viewed as the intensity of data coverage, at the space granularity of cells.

Remark 3.1 In some cases, it may not be necessary, or possible to count the exact number of data carriers in a cell, especially due to privacy concerns. For example, a carrying entity, who agrees to share data (participate in the dissemination) anonymously, may be reluctant to reveal its carrying status to others. For these cases, data strength can be measured from the perspective of recipients, that is, the number of datum δ an entity can receive in a cell during a time step. In other words, the system operator can employ probe entities to measure data strength.

Remark 3.2 In theory, any form of cell is acceptable. In fact, cells don't even have to be mutually exclusive, because as long as set $\mathcal{A} = \{A_1, A_2, \cdots, A_N\}$ jointly cover region A, vector $\mathbf{s}(\delta, t)$ will be a numerical description of data coverage $C(\delta, t)$ over the entire region A. From the data owner's perspective, it is also sensible to include all the sub-regions of interest as the collection $\mathcal{A}$, e.g., entrances, elevators and stairways of a mall in the mobile ad application. We adopt the partition of square cells for its simplicity in denotation and simulation.

When the area of a cell is larger than, or comparable to, the area of individual coverage, data-strength of cell A_n can be further simplified[1] to

$$s^n(\delta, t) := \sum_{e \in E} \delta_e(t) \mathbb{1}_{A_n}(p_e(t)). \tag{3.4}$$

In this way, data coverage, a time-varying subset of A, is transformed into a numeric data-strength vector, whose value can be calculated with attributes $\{\delta_\cdot(t), p_\cdot(t)\}_{e \in E}$ of all the entities. Next, we discuss the short-term evolution of the coverage dynamics, that is, how data strength $\mathbf{s}(\delta, t)$ over the region A changes over time, from the entity-level and system level, respectively.

[1] Entities in cell A_n can also obtain copies of δ from entities in adjacent cells. This part of data-strength is 'lost' due to partition, which is negligible when cells are large (compared to individual coverage of entities). The impact of cross-cell data forwarding is considered in Sect. 3.3.

3.3.2 State Transitions of a Single Entity

As can be seen in the state transition diagram (grey box in Fig. 3.1) of an entity, individual state changes are driven by the decision an entity makes. Nonetheless, these decisions are impossible for a large system to determine at an individual level. Therefore, we characterize their preferences on the data dissemination process at a system-level, with participating rate β and withdrawing rate γ.

Participating Rate $\beta \in [0, 1]$ is the probability that an arbitrary entity $e \in E$ chooses to *participate* during time step t, given that e has received one copy of datum δ during time step $t-1$. In other words, for an entity $e|_{\delta_e(t-1)=0,\ p_e(t-1)\in A_n}$ located in cell A_n at time $t-1$, the probability that its state changes from 0 to 1 during time step t can be obtained as

$$\begin{aligned}&\mathbb{P}\Big(\delta_e(t)=1 \;\Big|\; \{\delta_e(t-1)=0,\, p_e(t-1)\in A_n\}\Big)\\&\simeq 1-(1-\beta)^{f(\mathbf{s}(\delta,t))+\sum_{x\in E}\delta_x(t-1)B_{x,e}},\end{aligned} \tag{3.5}$$

where $f(\mathbf{s}(\delta,t)) = s^n(\delta,t-1)+c\sum_{j\in N(n)} s^j(\delta,t-1)$, in which c is a coefficient and $N(n)$ denotes the indices of cells that are adjacent to cell A_n; and quantity $B_{x,e}=1$ if and only if entity e is a stationary entity, and is connected to another stationary entity x via a wired link. In fact, the exponent $f(\mathbf{s}(\delta,t))$ in Eq. (3.5) equals to the number of datum copies that entity e expects to receive over wireless links during time step $t-1$. This includes datum copies from co-locating entities (inside cell A_n) and those from adjacent cells (only a fraction c of datum copies sent from entities that are close to cell A_i can be heard by e). For mobile entities, the last term in the exponent $f(\mathbf{s}(\delta,t))$ of Eq. (3.5) equals to 0.

For the clarity of description, we refer to the heterogeneous scenario as *Case-2*, which includes both wired and wireless connections, and the simpler scenario with merely wireless connections as *Case-1*, throughout this chapter.

Withdrawing Rate $\gamma \in [0, 1]$ is the ratio of participating entities that choose to stop disseminating data in the next time step, or equivalently, the probability that an entity $e|_{\delta_e(t-1)=1}$ takes a *withdraw* action (resulting in a state transition from 1 to 0), that is,

$$\mathbb{P}\big(\delta_e(t)=0 \mid \delta_e(t-1)=1\big)=\gamma,\ \forall t\in\mathcal{T}. \tag{3.6}$$

Remark 3.3 Note that the probability of participating increases with the total number of datum copies an entity receives, as revealed in Eq. (3.5), while the probability to withdraw is independent to actions of other entities, which can be seen from Eq. (3.6). The reason behind the increasing willingness to participate over the number of received data copies is that, more received copies indicate there are more participants in the system, which further implies a possibly high incentive offered by the data owner, hence the larger chance of participating. On the other hand, an entity already knows the incentive when it makes the decision on withdrawal, so its decision should not be affected by others' actions. Instead, an entity could withdraw from the dissemination process of datum δ due to lack of storage space, lost of network connectivity, or simply out of its own will, which are captured by probability γ.

As system-level indicators of an entity's willingness to participate, rate β and γ can be changed by adjusting the incentives offered to participants. By these rates, we are now able to link the individual states $\delta_e(t)$ to the data-strength $\mathbf{s}(\delta, t-1)$ of a single cell, via Eqs. (3.5) and (3.6). Next, we zoom out further to analyze the dynamics' evolution on the system level.

3.3.3 Evolution of the Dynamics via Data-Strength Vector $\mathbf{s}_t$

The key to understand the coverage dynamics $\{C(\delta, t)\}_{t\in\mathcal{T}}$ is knowing how it evolves in one time step, i.e., from $C(\delta, t-1)$ to $C(\delta, t)$, or equivalently under the data-strength measure, from vector $\mathbf{s}(\delta, t-1)$ to $\mathbf{s}(\delta, t)$. For the ease of notation, we omit δ, and write $\mathbf{s}(\delta, t)$ as $\mathbf{s}_t = [s_t^1, s_t^2, \cdots, s_t^N]^\mathsf{T}$, because we are only interested in the dissemination process of datum δ.

Consider cell $A_n \in \mathcal{A}$. The change of data-strength in this cell, $(\Delta\mathbf{s}_t)^n = s_t^n - s_{t-1}^n$, is induced by two impetuses, movements of entities (across cells), and data dissemination (inside cell A_n and across neighboring cells), which we discuss separately.

3.3.3.1 User Mobility Component

Let $p_t^n := \sum_{e\in E} \mathbb{1}_{A_n} p_e(t)$ denote the number of entities inside cell A_n at time t. We write its vector form as $\mathbf{p}_t = [p_t^1, p_t^2, \cdots, p_t^N]$. Let $m_{i,j}^t$ denote the number of entities that move out of cell A_i to a different cell A_j during time step t, i.e., $m_{i,j}^t = |\{e \in E | p_e(t-1) \times p_e(t) \in A_i \times A_j\}|$. For cell index $i \neq j$, define the i, j-th element of a $N \times N$ matrix $\mathbf{W}_t$, the user mobility matrix, as

$$W_{i,j}^t := \begin{cases} \frac{m_{i,j}^t}{p_t^n}, & \text{if } i \neq j \text{ and } p_t^n > 0, \\ 0, & \text{otherwise.} \end{cases} \tag{3.7}$$

Matrix $\mathbf{W}_t$ records the ratio of entities that move across cell boarders. During time step t, among the $W_{n,k}^t p_t^n$ entities that moved from cell A_n to cell A_k, on average, there are $(1-\gamma)W_{n,k}^t s_t^n$ data carriers, who contributes to the decrease in data-strength s_t^n, and the increase of s_t^k.

3.3.3.2 Data Dissemination Component

Assuming entities are 'well-mixed' in each cell,[2] we apply mean field approximation (MFA) at the cell level to derive the one-step evolution equation. We start from the simpler Case-1 with merely wireless connections. For cell A_n, on average, the number of new participants grows at rate $\beta f(\mathbf{s}(\delta, t))$ entities per time step, and existing participants withdraw at rate γs_{t-1}^n entities per time step.

With the entity-level state transition analysis and MFA, the average evolving trend of the coverage dynamics can be described with $\mathbf{W}_t$, $\mathbf{s}_t$ and $\mathbf{p}_t$. Specifically, for every cell A_n in a Case-1 scenario, the number of entities p_t^n and data-strength s_t^n satisfy

$$p_t^n = p_{t-1}^n + \sum_{j=1}^{N} W_{j,n}^t p_{t-1}^j - \sum_{k=1}^{N} W_{n,k}^t p_{t-1}^n, \tag{3.8}$$

$$\begin{aligned} s_t^n = {} & (1-\gamma)\left(s_{t-1}^n + \sum_{j=1}^{N} W_{j,n}^t s_{t-1}^j - \sum_{k=1}^{N} W_{n,k}^t s_{t-1}^n\right) \\ & + (p_{t-1}^n - s_{t-1}^n)\beta\left(s_{t-1}^n + c \sum_{m \in N(n)} s_{t-1}^m\right). \end{aligned} \tag{3.9}$$

A clear observation from Eq. (3.9) is that the signal value of a cell at time t, is the summation of a scaled version of its previous value, and the influence of its *neighboring* cells, which are captured by the user mobility matrix $\mathbf{W}_t$ and $N(n) \subset \mathcal{A}$. For the Case-2 scenario, the number of entities in each cell p_t^n still satisfies Eq. (3.8), while the data strength may increase due to data transmission along wired connections, which are time-invariant, and accessible before the data dissemination process. Denote $\mathrm{AP}_n := \{e \in E \mid p_0^e = p_1^e = \cdots = p_t^e \in A_n\}$ as the set of stationary entities in cell A_n, and $L_{n,j} := \sum_{e,x \in N^2} \mathbb{1}_{\mathrm{AP}_n}(e) B_{ex} \mathbb{1}_{\mathrm{AP}_j}(x)$ as the number of wired connections between cell A_n and A_j. Recall B_{ex} is defined as 1 when entity e and x are connected permanently by a wired link, in Eq. (3.5). Then data strength s_t^n in the Case-2 scenario has one more term than Eq. (3.9), i.e.,

$$\begin{aligned} s_t^n = {} & (1-\gamma)\left(s_{t-1}^n + \sum_{j=1}^{N} W_{j,n}^t s_{t-1}^j - \sum_{k=1}^{N} W_{n,k}^t s_{t-1}^n\right) \\ & + (p_{t-1}^n - s_{t-1}^n)\beta\left(s_{t-1}^n + c \sum_{m \in N(n)} s_{t-1}^m\right) + \frac{p_{t-1}^n - s_{t-1}^n}{p_{t-1}^n}\beta \sum_{j=1}^{N} L_{n,j} \frac{s_{t-1}^j}{p_{t-1}^j}. \end{aligned} \tag{3.10}$$

[2] 'Well-mixed' means each entity has an equal chance of contact (transmit/receive datum) with any other entities in the same cell. We set the cell size to be comparable to individual coverage (in terms of area), so that every entity in a cell is almost under the individual coverage of each other.

Equation (3.10) indicates that the time-invariant matrix $\mathbf{L}_{N\times N}$ adds another layer of interconnection between cells (in addition to the user mobility matrix $\mathbf{W}_t$), which is independent of geographical adjacency (as opposed to matrix $\mathbf{W}_t$). Together, Eqs. (3.8), (3.9), and (3.10) sketch the average evolution trend of the data-strength with the following implications.

1. *Mobility drives the coverage dynamics*. The change in data-strength during time step t, $\Delta\mathbf{s}_t = \mathbf{s}_t - \mathbf{s}_{t-1}$, relies heavily on user mobility (captured by matrix $\mathbf{W}_t$ in Eqs. (3.8) and (3.9)), which is to say, the change in the data strength of each cell can be written as a function of $\mathbf{W}_t$. This is especially true when the participating rate β (indicator of data forwarding) is small or the withdrawing rate γ is small, as both the last term in Eq. (3.9) and the extra term in Eq. (3.10) also tend to be small, compared to the first term in Eq. (3.9). In addition, term $\sum_{m\in N(n)} s_{t-1}^m$ in Eq. (3.9) can be re-written as a function of $\mathbf{W}_t$ and $\mathbf{s}_{t-1}$, as cross-cell movements (in one time step) are only possible, if the two cells are close to each other.
2. *Data-strength* $\mathbf{s}_t$ *is a* ***graph signal***. By nature, any data-strength vector $\mathbf{s}_t$ is essentially a vector of positive real numbers indexed by cells, i.e., a *signal* defined on a graph with vertex set $\mathcal{A}$. Given an initial state $\mathbf{s}_0$, the signal value s_t^n of cell A_n can be estimated with the data-strength in the last time step, i.e., $\mathbf{s}_{t-1}$ and matrix $\mathbf{W}_t$, as shown in Eq. (3.9). In fact, only a few neighboring cells (determined by elements of $\mathbf{W}_t$) are relevant to the change of data-strength during time step t, because correlation between signal values s_t^n and s_{t-1}^m decreases as the distance between cell A_n and A_m increases, which is true for both Eqs. (3.8) and (3.9). Therefore, matrix $\mathbf{W}_t$ is closely related to the underlying structure of signal $\mathbf{s}_t$, and can be viewed as the adjacency matrix of the graph. Formulating $\mathbf{s}_t$ as a graph signal allows us to explore the rich property of coverage dynamics in both the space domain and the time domain, as a preparation to which, we briefly introduce the basics of graph signal processing (GSP) for self-containment reasons.

3.3.4 Preliminaries on Graph Signal Processing (GSP)

Graph signal processing (GSP) is an emerging mathematical tool to describe, analyze, and recover high-dimensional data. On this avenue, Sandryhaila and Moura extended digital signal processing (DSP) concepts to static signals index by graphs, and formally defined the graph shift operator, graph Fourier transform (GFT) and graph filters based on adjacency matrices [17]. Similar concepts are also defined based on the graph Laplacian [18], which has nice properties such as positive semi-definite, to make GSP more commutable with DSP. Then taking the evolution of graph signals into consideration, Grassi et al. constructed a time-vertex signal processing framework, which formally defined time-vertex process, joint time-vertex Fourier transform (JFT) [9], and filter banks [5]. Based on this framework, prediction of stationary graph signal with frequency domain analysis is made possible in [9], and proved to have a lower complexity. Recently, Marques et al. generalized the definition of weak stationarity for random graph signals, and proposed a spectral estimation mechanism to predict real-world graph signals [11].

Next, we briefly introduce terminologies in GSP that are useful in this chapter. Interested users are readers are directed to [5, 9, 11, 17, 18] for details.

3.3.4.1 Graph Fourier Transform (GFT) and Spectrum of a Graph Signal

Consider graph signal $\mathbf{s}_t \in \mathbb{R}^N$ on an underícted and connected graph $\mathcal{G}(\mathcal{A}, \mathbf{W})$, in which $\mathcal{A}$ is the vertex set, and $\mathbf{W}$ is the weighted adjacency matrix of $\mathcal{G}$. Let $\mathbf{D}$ be the generalized degree matrix of $\mathcal{G}$, that is, a diagonal matrix with diagonal elements $\mathbf{D}_{i,i} = \sum_{j\in\mathcal{A}} w_{i,j}$. Then the **Laplacian** of graph $\mathcal{G}$ is defined as $\mathcal{L}_{\mathcal{G}} := \mathbf{D} - \mathbf{W}$. Since $\mathcal{G}$ is undirected, $\mathcal{L}_{\mathcal{G}}$ is real-valued, symmetric and positive semi-definite, as a result of which, $\mathcal{L}_{\mathcal{G}}$ has N real, non-negative eigenvalues $\{\lambda_n\}_{n=0,1,\ldots,N-1}$.

Without loss of generality, let the eigenvalues $\lambda_0 \leq \lambda_1 \leq \cdots \leq \lambda_{N-1}$ be non-decreasingly ordered, and denote $\Lambda = diag(\lambda_0, \lambda_1, \cdots, \lambda_N)$. Then we always have $\lambda_0 = 0$, and each eigenvalue λ_n corresponds to an eigenvector $\mathbf{v}_n$ (column vector). Let $\mathbf{U}_{\mathcal{G}} = [\mathbf{v}_0, \mathbf{v}_1, \cdots, \mathbf{v}_{N-1}]$ be the matrix form of all the eigenvectors of $\mathcal{L}_{\mathcal{G}}$. Then graph Laplacian $\mathcal{L}_{\mathcal{G}} = \mathbf{U}_{\mathcal{G}}\Lambda\mathbf{U}_{\mathcal{G}}^*$. The **GFT** of signal $\mathbf{s}_t$ is defined as the product of $\mathbf{U}_{\mathcal{G}}$ and the graph signal $\mathbf{s}_t$, that is,

$$\widehat{\mathbf{s}_t} = \mathrm{GFT}\{\mathbf{s}_t\} := \mathbf{U}_{\mathcal{G}}^*\mathbf{s}_t, \tag{3.11}$$

where $\mathbf{U}_{\mathcal{G}}^*$ is the conjugate transpose of $\mathbf{U}_{\mathcal{G}}$, and

$$\widehat{\mathbf{s}_t}_{N\times 1} = [\mathbf{s}_t(\lambda_0), \mathbf{s}_t(\lambda_1), \cdots, \mathbf{s}_t(\lambda_{N-1})]^{\mathsf{T}} \tag{3.12}$$

is the GFT spectrum of $\mathbf{s}_t$. Like the spectrum of a time-varying 1-d signal, GFT spectrum $\widehat{\mathbf{s}_t}$ of graph signal $\mathbf{s}_t$ captures the *variation* of the signal in the vertex domain, which in our case contains all the cells in $\mathcal{A}$.

3.3.4.2 Time-Vertex Process and Stationarity

Temporal evolution of a graph signal composes a time-vertex process $\mathbf{S}_{N\times T} = [\mathbf{s}_1; \mathbf{s}_2; \cdots; \mathbf{s}_T]$. Along the time dimension, every row vector of $\mathbf{S}$, can be analyzed through discrete Fourier transform (DFT). The DFT of process $\mathbf{S}$ is defined as $\mathrm{DFT}\{\mathbf{S}\} := \mathbf{S}\overline{\mathbf{U}_T}$, where $\mathbf{U}_T$ is the T-point normalized DFT matrix. Its (t, s)-element is $\mathbf{U}_T(t, s) = \frac{e^{-j\omega_s t}}{\sqrt{T}}$, and $\omega_s = \frac{2\pi(s-1)}{T}$.

Recall in Sect. 3.3.3, we find that user mobility drives the evolution of data coverage dynamics. In regard to user mobility and data dissemination, Grossglauser and Tse proved that mobility increases per-user throughput in ad hoc wireless networks [6], when mobile nodes function as relays. Naturally, we anticipate data coverage to be boosted by an increase in mobility (speed), because: (1) In both Case-1 and Case-2 scenario, entities function as relay, passing the datum to others. (2) Increased mobility leads to an increase of probability that a relay can meet with a destination (in this case, any entity), as well as pair-wise throughput [6], which indicates that more data transmissions between entities can happen simultaneously. But an open question is, *how much* does mobility (captured by $\mathbf{W}_t$) and data forwarding (captured by β) contribute to the change of data-strength $\Delta\mathbf{s}_t$?

3.4 Information from a Snapshot

Knowing quantitatively how much entity mobility and data forwarding contribute to the dynamic evolution of data coverage can significantly benefit data disseminators, who expect to boost data coverage effectively. For example, which works better for an ad distributor: raise incentive to increase the participating probability β, or employ faster-moving entities as distributors? This is especially meaningful to the Case-1 scenario, in which wired connections are not available for data dissemination.

Therefore, in this section, we discuss the impact of user mobility in the Case-1 scenario. To be more specific, we examine a snapshot of the coverage dynamics, i.e., the difference graph signal $\Delta\mathbf{s}_t = \mathbf{s}_t - \mathbf{s}_{t-1}$, with respect to user mobility $\mathbf{W}_t$. A simple homogeneous case with a synthetic mobility model is considered in this section, to better uncover the impetuses.

3.4.1 A Simple Homogeneous Scenario

Suppose the entire network is composed of mobile entities that (1) have the same transmission range r_c, and (2) move according to a speed-constrained mobility model detailed as follows. During the 'move' stage in every time step, each entity $e \in E$ chooses a direction uniformly at random from $[0, 2\pi)$, along which e moves at a speed randomly chosen from $[0, v_{max}]$. We follow the convention in [7] and assume that, an entity crossing the boundary of region A on one side will emerge from the opposite side. In addition, considering that every entity withdraws independently, we set the withdrawal rate $\gamma = 0$ in this section to focus on the impact of maximum speed v_{max}, participating rate β, and communication range r_c.

3.4.2 Impact of Mobility

Recall data-strength vector $\mathbf{s}_t$ (similarly the difference signal $\Delta\mathbf{s}_t$) is a graph signal defined on vertex set $\mathcal{A}$ of some graph $\mathcal{G}$. Also, we observed the important role of user mobility matrix $\mathbf{W}_t$ in the dynamical change of data strength, in Eq's. (3.8), (3.9), and (3.10), so we propose to define the weighted adjacency matrix $\widetilde{\mathbf{W}}_t$ of graph $\mathcal{G}$ as a function of $\mathbf{W}_t$, which is fairly accessible, due to prevalent GPS devices [15], IoT sensors [16], and location-based services [19].

3.4.2.1 Weighted Adjacency Matrix $\widetilde{\mathbf{W}}_t$ of $\mathcal{G}$

The evolution rule in Eq. (3.9) is formulated based on the instantaneous user mobility $\mathbf{W}_t$, but it is not suitable to directly assign $\mathbf{W}_t$ as the adjacency matrix of $\mathcal{G}$ for real systems, because of the time gap between data transmission (several ms to seconds) and positioning (e.g., GPS logging, seconds to minutes). To bridge this gap, we observe the system every $K \in \mathbb{N}^+$ time steps, and obtain a snapshot $\Delta\mathbf{s}_t^K = \mathbf{s}_t - \mathbf{s}_{(t-K+1)}$ for every observation. Accordingly, we employ the symmetric accumulated version $\widetilde{\mathbf{W}}_t$ as the weighted adjacency matrix of graph $\mathcal{G}$, defined as the number of cell crossings during time interval of length K.

Remark 3.4 We use the number of crossings, instead of ratio of crossing in Eq. (3.7), due to two reasons: (1) it is easier to obtain offline for a period of time, and (2) the number of cell crossings also captures correlation between adjacent cells ($N(n)$ in Eqs. (3.9) and (3.10)), because the more cell crossing, the more entities gathering at the cell edge, and hence a higher probability of data forwarding across cell boarders. In this way, we are able to reduce the number of variables in Eqs. (3.9) and (3.10), and write $\mathbf{s}_t = \left[(1-\gamma+\beta)^K\mathbf{I} + f(t)\widetilde{\mathbf{W}}_t\right]\mathbf{s}_{t-K+1} + \boldsymbol{\psi}_t$ instead, where $f(t)$ is a coefficient function by β and γ, and $\boldsymbol{\psi}_t$ is the error term introduced by the approximation.

To validate the graph formation based on $\widetilde{\mathbf{W}}_t$, and the speculation that the change in data-strength, $\Delta\mathbf{s}_t^K$, is a graph signal on $\mathcal{G}$, we first examine three snapshot instances (with the same user mobility and initial state, so graph $\mathcal{G}$ and the weight matrix $\widetilde{\mathbf{W}}_t$ are exactly the same). As can be seen from the GFT spectrum in Fig. 3.7, high participating rate β will induce peaks at high frequencies, that is, larger products $\widehat{\Delta\mathbf{s}}(\lambda_n) =< \Delta\mathbf{s}, \mathbf{v}_n >$ at larger eigenvalues λ_n of the graph Laplacian $\mathcal{L}_\mathcal{G}$ of $\mathcal{G}$. This conceptually validates our observation of $\Delta\mathbf{s}_t^K$ being a graph signal on $\mathcal{G}$, which is defined by matrix $\widetilde{\mathbf{W}}_t$. However, spectrum $\widehat{\Delta\mathbf{s}}$ of signal $\Delta\mathbf{s}_t^K$ does not intuitively or quantitatively reflect the impact of user mobility. Therefore, we define a new metric to quantify the impact.

3.4.2.2 Mobility Dependence Index (MDI)

With user mobility quantitatively represented by $\widetilde{\mathbf{W}}_t$, the level of *inconsistency* of difference signal $\Delta\mathbf{s}_t$ with respective to $\mathcal{G}$ reflects the impetus that is not from mobility, i.e., from data forwarding, as illustrated in Fig. 3.7. To quantify the impact of mobility, we define a new metric for every interval $[t, t+K-1]$:

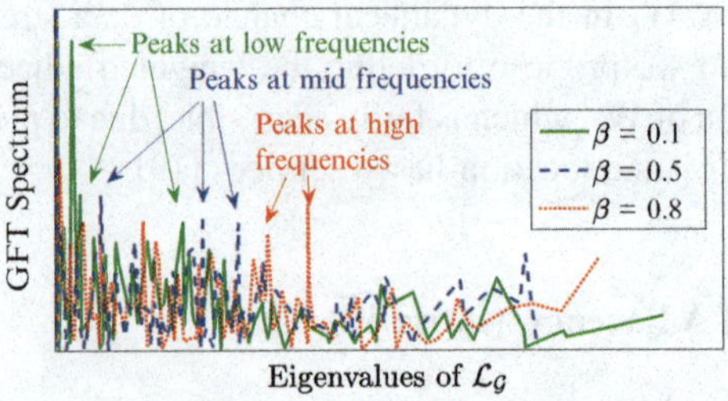

Fig. 3.7 GFT spectrum of three difference signals $\Delta\mathbf{s}_t$, generated with different β values, illustrate the variation of $\Delta\mathbf{s}_t$ over mobility $\widetilde{\mathbf{W}}_t$: the higher the participating probability β, the more dominant the high frequency components, and the less impact by user mobility

Definition 3.2 The **mobility dependency index** of a snapshot $\Delta\mathbf{s}_t^K$ with respective to $\widetilde{\mathbf{W}}_t$ is defined as

$$\mathrm{MDI}_{\widetilde{\mathbf{W}}_t}(\Delta\mathbf{s}_t^K) := \frac{1}{\|\Delta\mathbf{s}_t^K - \frac{1}{\lambda_{max}}\widetilde{\mathbf{W}}_t\Delta\mathbf{s}_t^K\|_1}, \tag{3.13}$$

where λ_{max} is the maximum eigenvalue of $\widetilde{\mathbf{W}}_t$.

In this definition, the denominator is the *total variation* $\mathrm{TV}_{\mathcal{G}}$ of the difference signal $\Delta\mathbf{s}_t^K$. Intuitively, $\mathrm{TV}_{\mathcal{G}}$ quantifies the conformity of $\Delta\mathbf{s}_t^K$ with respect to entity movements captured by weight matrix $\widetilde{\mathbf{W}}_t$. Therefore, the higher the MDI (i.e., lower $\mathrm{TV}_{\mathcal{G}}$), the more mobility contributes to the evolution of coverage dynamics. When the difference signal $\Delta\mathbf{s}_t^K$ is very smooth (with respect to graph $\mathcal{G}$), the change in data-strength is aligned with weight matrix $\widetilde{\mathbf{W}}_t$, indicating a heavy influence of mobility. In contrast, any increment in signal value attributed to data forwarding will induce an increment in $\mathrm{TV}_{\mathcal{G}}$, and hence a decrease in the MDI value.

3.4.2.3 Simulation Configuration

To observe the impact of user mobility via the MDI metric, we simulate a dissemination process in the wireless connection only Case-1 scenario, whose configurations shown in Table 3.1.

3.4.2.4 Observations

Figures 3.8 and 3.9 present the MDI value of the difference signal $\Delta\mathbf{s}_t^K|_{K=5}$ generated with the same initial condition $\mathbf{s}_t$, but with different participating rate β, maximum speed v_{max} and transmission range r_c. We highlight the following observations.

Table 3.1 Simulation configuration

Para.	Description	Value	Para.	Description	Value
A	Region of interest	$[0, 1000]^2$	$\|E\|$	# of entities	1000
$\|A_i\|$	Area of a cell	100×100	N	# of cells in A	100
r_c	Communication range	[10, 50]	v_{max}	Max. speed	[10, 100]
N/A	# of seeds	50	N/A	# of runs	100
β	Participating prob.	[0.05, 0.4]	γ	Withdrawing prob.	0
Δ	Time step	1 s	K	Observation interval	5 s

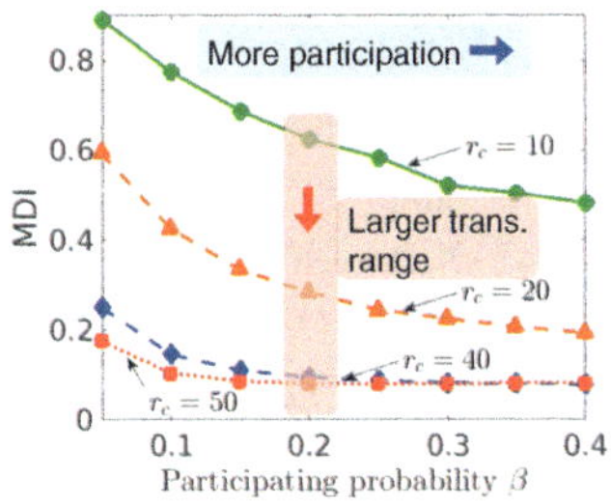

Fig. 3.8 MDI vs. β ($v_{max} = 10$): Bounded impact of data forwarding on the evolution of data coverage dynamics

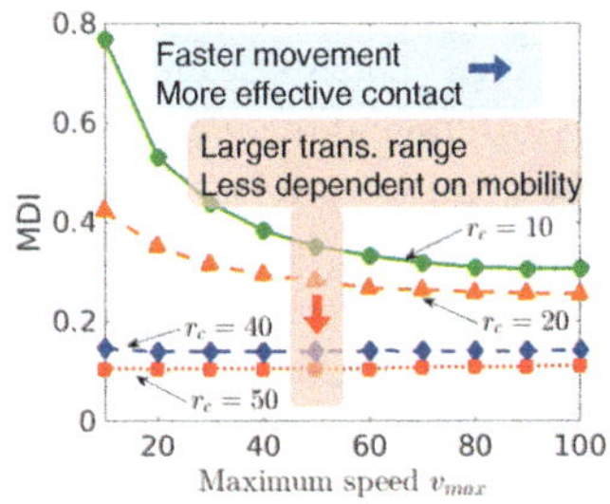

Fig. 3.9 MDI vs. maximum speed v_{maxf}: Bounded impact of entity mobility on the evolution of data coverage dynamics

1. *Trend of MDI*: It is intuitive that MDI decreases when transmission range r_c increases (Figs. 3.8 and 3.9), and also decreases when the participating probability β increases (Fig. 3.8), because increment in either r_c or β will result in more non-carrying entities ($\delta_e(t) = 0$) to receive a copy of δ and participate in the dissemination. Nonetheless, it is rather interesting to observe in Fig. 3.9, that when the maximum speed v_{max} increases, MDI decreases dramatically for small transmission ranges. The reasons behind this plunge are: (1) the increment in data-strength largely relies on new effective contacts, that is, data forwarding between entity-pairs, which are more prevalent due to the larger movement range ($K v_{max}$); and (2) elements in user mobility matrix $\widetilde{\mathbf{W}}_t$ can not capture geographical adjacency of cells any more, as movements between non-adjacent cells are now possible due to the larger movement range.
2. *Decreasing, but bounded impact*: observe that MDI flats out as β and v_{max} further increases, indicating a *threshold behavior* of the coverage dynamics with respect to both impetuses: the change of data coverage can not always be accelerated by increasing β or v_{max}. In fact, above a certain threshold, e.g., $\beta = 0.2$ at communication range $r_c = 40$, further increasing β will not decrease the MDI in

Fig. 3.8, so β has bounded impact on the coverage change. Similar effect is also seen for the maximum speed v_{max}.
3. *Prediction based on* $\widetilde{\mathbf{W}}_t$: observe the high MDI when maximum speed v_{max} (and preferably the participating rate β as well) of entities is relatively low, which is also captured by Eqs. (3.9) and (3.10). This phenomenon indicates that it is possible to predict data coverage based on the weight matrix $\widetilde{\mathbf{W}}_t$, which is quite accessible in a real-world edge network.

3.5 Summary

In this chapter, we raise the question of *where data are*, in heterogeneous edge networks, which is becoming the main scenario for data dissemination services. To address the *whereabouts* of mobile data, we formally define the scope, entity model, and data coverage for a data dissemination process in such networks, propose a data-strength metric such that data coverage can be quantified, and formulate the data coverage dynamics as a time-vertex process of numeric graph signals. This formulation enables us to observe, analyze, and predict the dynamic evolution of coverage dynamics, en route of data movements. Specifically, from snapshots of the coverage dynamics, we observe the impact of user mobility is high, especially when entities are less active forwarding or the maximum moving speed is low; as the maximum speed increases, such impact decreases but is still lower bounded. From this observation, we infer that it is possible to predict data coverage based on system-level user mobility traces, which can be easily obtained without raising privacy concerns.

References

1. Andrienko, G., Andrienko, N., Chen, W., Maciejewski, R., Zhao, Y.: Visual analytics of mobility and transportation: State of the art and further research directions. IEEE Trans. Intell. Trans. Syst. **18**(8), 2232–2249 (2017). https://doi.org/10.1109/TITS.2017.2683539
2. Bilal, K., Khalid, O., Erbad, A., Khan, S.U.: Potentials, trends, and prospects in edge technologies: fog, cloudlet, mobile edge, and micro data centers. Comput. Netw. **130**, 94 – 120 (2018). https://doi.org/10.1016/j.comnet.2017.10.002. http://www.sciencedirect.com/science/article/pii/S1389128617303778
3. Cisco Visual Networking Index: Forecast and Trends, 2017–2022 White Paper. Technical Report, Cisco and/or its affliates (2019)
4. Felt, M.: Social media and the social sciences: How researchers employ big data analytics. Big Data Soc. **3**(1), 2053951716645828 (2016). https://doi.org/10.1177/2053951716645828. http://dx.doi.org/10.1177/2053951716645828
5. Grassi, F., Loukas, A., Perraudin, N., Ricaud, B.: A time-vertex signal processing framework: scalable processing and meaningful representations for time-series on graphs. Trans. Sig. Proc. **66**(3), 817–829 (2018). https://doi.org/10.1109/TSP.2017.2775589
6. Grossglauser, M., Tse, D.N.C.: Mobility increases the capacity of ad hoc wireless networks. IEEE/ACM Trans. Netw. **10**(4), 477–486 (2002). https://doi.org/10.1109/TNET.2002.801403

7. Li, Y., Wang, W.: Horizon on the move: Geocast in intermittently connected vehicular ad hoc networks. In: 2013 Proceedings IEEE INFOCOM, pp. 2553–2561 (2013). https://doi.org/10.1109/INFCOM.2013.6567062
8. Li, Z., Wang, C., Yang, S., Jiang, C., Stojmenovic, I.: Space-crossing: community-based data forwarding in mobile social networks under the hybrid communication architecture. IEEE Trans. Wirel. Commun. **14**(9), 4720–4727 (2015). https://doi.org/10.1109/TWC.2015.2424965
9. Loukas, A., Isufi, E., Perraudin, N.: Predicting the evolution of stationary graph signals. In: 2017 51st Asilomar Conference on Signals, Systems, and Computers, pp. 60–64 (2017). https://doi.org/10.1109/ACSSC.2017.8335136
10. Lu, Z., Wang, W., Wang, C.: How can botnets cause storms? understanding the evolution and impact of mobile botnets. In: INFOCOM, 2014 Proceedings IEEE, pp. 1501–1509 (2014). https://doi.org/10.1109/INFOCOM.2014.6848085
11. Marques, A.G., Segarra, S., Leus, G., Ribeiro, A.: Stationary graph processes and spectral estimation. IEEE Trans. Sig. Process. **65**(22), 5911–5926 (2017). https://doi.org/10.1109/TSP.2017.2739099
12. Ning, H.: Unit and Ubiquitous Internet of Things. CRC Press, Boca Raton (2013). https://books.google.com/books?id=5csnAAAAQBAJ
13. Qin, J., Zhu, H., Zhu, Y., Lu, L., Xue, G., Li, M.: Post: Exploiting dynamic sociality for mobile advertising in vehicular networks. In: IEEE INFOCOM 2014 - IEEE Conference on Computer Communications, pp. 1761–1769 (2014). https://doi.org/10.1109/INFOCOM.2014.6848114
14. Qin, J., Zhu, H., Zhu, Y., Lu, L., Xue, G., Li, M.: Post: Exploiting dynamic sociality for mobile advertising in vehicular networks. IEEE Trans. Parall. Distr. Syst. **27**(6), 1770–1782 (2016). https://doi.org/10.1109/TPDS.2015.2467392
15. Research, G.V.: Global Positioning Systems (GPS) Market Size, Share & Trends Analysis Report By Deployment, By Application (Aviation, Marine, Surveying, Location-Based Services, Road), And Segment Forecasts, 2018 - 2025. Technical Report, Grand View Research (2018)
16. Safavi, S., Khan, U.A., Kar, S., Moura, J.M.F.: Distributed localization: a linear theory. Proc. IEEE **106**(7), 1204–1223 (2018). https://doi.org/10.1109/JPROC.2018.2823638
17. Sandryhaila, A., Moura, J.M.F.: Discrete signal processing on graphs: frequency analysis. IEEE Trans. Sig. Process. **62**(12), 3042–3054 (2014). https://doi.org/10.1109/TSP.2014.2321121
18. Shuman, D.I., Narang, S.K., Frossard, P., Ortega, A., Vandergheynst, P.: The emerging field of signal processing on graphs: Extending high-dimensional data analysis to networks and other irregular domains. IEEE Sig. Process. Mag. **30**(3), 83–98 (2013)
19. Steinfield, C.: The development of location based services in mobile commerce, pp. 177–197. Physica-Verlag HD, Heidelberg (2004). https://doi.org/10.1007/978-3-662-11659-3_10
20. Tao, F., Qi, Q., Liu, A., Kusiak, A.: Data-driven smart manufacturing. J. Manuf. Syst. **48**, 157 – 169 (2018). https://doi.org/10.1016/j.jmsy.2018.01.006. http://www.sciencedirect.com/science/article/pii/S0278612518300062. Special Issue on Smart Manufacturing
21. Wang, X., Wu, W., Qi, D.: Mobility-aware participant recruitment for vehicle-based mobile crowdsensing. IEEE Trans. Veh. Technol. **67**(5), 4415–4426 (2018). https://doi.org/10.1109/TVT.2017.2787750
22. Wedel, M., Kannan, P.: Marketing analytics for data-rich environments. J. Market. **80**(6), 97–121 (2016). https://doi.org/10.1509/jm.15.0413
23. Yan, T., Zhang, W., Wang, G.: Dove: Data dissemination to a desired number of receivers in vanet. IEEE Trans. Veh. Technol. **63**(4), 1903–1916 (2014). https://doi.org/10.1109/TVT.2013.2287692
24. Zhang, H., Zhang, Z., Dai, H.: Gossip-based information spreading in mobile networks. IEEE Trans. Wirel. Commun. **12**(11), 5918–5928 (2013). https://doi.org/10.1109/TWC.2013.100113.130619
25. Zhang, K., Leng, S., Peng, X., Pan, L., Maharjan, S., Zhang, Y.: Artificial intelligence inspired transmission scheduling in cognitive vehicular communications and networks. IEEE Internet Things J. 1–1 (2019). https://doi.org/10.1109/JIOT.2018.2872013

Chapter 4
Governing Rules: Modeling and Analysis of Task Offloading Processes in the Fog

In the previous chapters, we discussed the dissemination process of *one* data block, particularly its lifetime and whereabouts, in wireless networks. As data services quickly migrate to the network edge, which is composed of numerous resource-constrained wireless devices and access network elements, the dissemination processes of *multiple* data blocks may interfere with each other, under the competition for resources. A typical example is the emerging *fog* paradigm, in which a *task*, that is, a computation-intensive (and delay-sensitive) data block, can be offloaded to resourceful fog nodes nearby, in order to reduce latency, or save energy. The offloading decision is usually made by considering the amount of accessible resource in the network, while its execution in turn affects how other tasks will be offloaded, rendering it difficult to evaluate the performance of such large-scale fog systems.

To address this open problem, we study the task offloading process in terms of *how multiple data blocks move* in this chapter. We propose a gravity-based offloading model, to describe how the target fog node is selected under different offloading criteria. This model allows us to evaluate the performance of a large-scale fog system, particularly how much (local and shared) resource such offloading processes consume, which are quantified by the *device effort* and *network effort* metrics. With respect to the scalability of fog, which is a major design concern of the paradigm, we find that, under the gravity-based offloading rule, the lifetime of an individual task decreases as the size of the network increases, while the total resource needed for the system scales linearly. As our model can describe various offloading criteria, these results are generally applicable in understanding how data move in a fog network, and the amount of resource needed for such systems.

J. Wang et al., *Encountering Mobile Data Dynamics in Heterogeneous Wireless Networks*, https://doi.org/10.1007/978-3-031-62906-8_4

4.1 Motivation and Related Work

With the proliferation of smart devices [12, 31], including smart phones, wearables, and sensors, Internet-of-Things (IoT) is driving a digital transformation in all aspects of modern life [1]. As the vision quickly turns into reality [22], IoT brings about new challenges to its provision network: stringent latency requirement, resource-limited devices, and the prohibitive scale. These requirements are difficult to satisfy with the existing centralized cloud computing paradigm, where cloud servers are usually deployed in privately-owned, remote data centers. Consequently, *fog computing* is introduced [31] and envisioned as a promising paradigm in IoT provisioning [1].

4.1.1 Fog Emerges on the Edge: Remedy or Resource Drain?

The defining feature of the fog paradigm is bringing data services to the *edge* of the network, such that the processing and dissemination of data are available right where the generation and consumption of data take place. Instead of drawing the surging mobile data traffic to cloud servers through the busy core network, data are processed by (less powerful) fog servers deployed in edge devices, e.g., commercial edge routers, Cloudlet and IOx [39], and even mobile devices [15], which are referred to as *fog nodes*. In this sense, fog computing is a broad concept that includes mobile edge computing (MEC), mobile cloud computing (MCC), cloudlet, and mobile ad hoc cloud computing (MACC) [40]. From the perspective of network architecture, some researchers view fog nodes as an intermediate layer between "things" in the IoT and remote servers in the cloud [15, 17], while others view it as a fully distributed substitute of the cloud architecture [11, 30]. Despite these differences, a consensus for the fog continuum [31] is that, its monumental purpose is to provide data offloading options, with two design objectives [17] tailored for IoT applications:

1. *Short Delay*. Per request of real-time applications in IoT, such as smart infrastructure, autonomous driving, and virtual assistance [22], a fog system is expected to reduce the service latency to the ms level (< 10 ms), comparing to the 30$\sim$100 ms latency in the current cloud computing paradigm. This demand requires the fog paradigm to maintain a sufficient number of fog servers in the vicinity of any *task node*, which can be any wireless device seeking to offload data.
2. *Scalability*. Due to the massive scale of the geo-dispersed IoT systems [1], a fog system is expected to process tasks offloaded from a large number of task nodes, e.g., the vast amount of sensors, cameras, and smart vehicles in an intelligent transportation systems (ITS). Considering frequent changes of locations, connectivity and offloading criteria, this objective requires the fog paradigm to scale easily, i.e., operating with simple organization and minimum

overhead. In fact, openness and scalability have been recognized as the major design concerns, as pointed out by the OpenFog consortium in the IEEE OpenFog reference architecture [17].

With these objectives fulfilled, the fog paradigm seems a good remedy for the large-scale delay-sensitive IoT applications, as it rallies up resource at the network edge, and takes advantage of the huge numbers of candidate fog nodes (routers, AP, and mobile devices) in a distributed manner. However, as mobile traffic continues to grow [12], will the resource-limited network edge be able to keep such a fog eco-system afloat? In other words, *how to evaluate the performance of the fog paradigm* with respect to the massive task offloading processes?

Answer to this question is essential to understand the impact of mobile data services, which is non-trivial for the edge network, a heterogeneous wireless system. First, from the perspective of service provisioning, resource constraint has been a major issue at the network edge [21], especially for communication resources. While processing and storage can be boosted by simply deploying more fog nodes, communication resource, including spectrum, time, code, and space, is *shared* by the system, and hence subject to further degeneration as the system scales. Nonetheless, fog computing allows power-constrained end devices to save energy by avoiding local processing, at the cost of frequent short-range communications with nearby fog nodes, which further aggravates the spectrum scarcity concern [8]. Second, from the perspective of traffic patterns, IoT applications are so broad and complex, that data are generated in various volume, form, and frequency. For example, there are periodic, short sensory data from smart meters, as well as spontaneous, long video clips from ground traffic monitoring. As a result, the fog paradigm, which provides infrastructure-as-a-service (IaaS) and platform-as-a-service (PaaS) to IoT, impose unpredictable resource demand on the underlying edge network. For such a large-scale system that is still in the design stage, it is essential to understand its impact on the existing system, and its performance as the system scales.

4.1.2 Related Work

As an exemplified application in the fog, *data offloading* has been extensively studied in MCC, MEC, and cloudlets, from aspects of synchronized data storage [33], bench-marking [15], deployment [26], and so on. As fog is specifically designed for IoT applications, in which data blocks are generated by resource-limited nodes, and need to be processed within a time bound, we focus on the offloading process of computation-intensive data blocks [40], i.e., *tasks*.

The central question in task offloading is, whether a task should be offloaded, and if so, to which fog node? Answers to both questions are determined by the offloading mechanism. Existing literature on this topic can be categorized into centralized and distributed schemes, typical examples of which are shown in Table 4.1.

The former family of centralized schemes relies on a single central entity to make all the offloading decisions, e.g., [10], which to some extent is against the open nature of fog computing. The major challenge in this setting is offloading/scheduling decision with partial system information, to address which reinforcement learning(RL)-based schemes [4–6, 13, 27, 38] are recently developed. These schemes are usually designed for homogeneous systems of small scales, in the sense that task nodes adopt the same radio access technology (RAT), and cooperate to minimize the average task completion time, or the total energy consumption of the system. Nonetheless, real-world IoT systems are often-times large-scale and hence much more diverse, in aspects of RAT, task types, and offloading criteria.

The distributed approaches further branch into cooperative and non-cooperative types. In cooperative offloading schemes, e.g., [24, 37, 41], fog nodes exchange information to optimize resource allocation, or collectively schedule the offloaded tasks, usually to minimize the offloading latency [36] and/or energy consumption [35]. In other words, fog nodes cooperate to determine the offloading target for each task. Though cooperation among fog nodes is highly beneficial to QoS provisioning and load-balancing, it is less likely to perform cooperation in a large-scale fog continuum, due to the prohibitive communication overhead, lack of inter-operability among different providers, and security/privacy concerns of end users.

In this regard, Tang and He recognized the competition among fog nodes, modeled the offloading process as a non-cooperative game [29], and evaluated their solution with small-scale experiments. With a similar game theoretic approach, Jošilo and Dán modeled task allocation as a non-cooperative game, and proposed decentralized random offloading scheme, which requires the first and second moment of task attributes, but achieves nearly the same task completion time as myopic best response solutions that require real-time node-specific information [18]. Considering wireless links may not be stable during offloading, Wu studied fault recovery in the offloading process, and aimed to maximize revenue of the network [34].

Table 4.1 summarizes the characteristics and parameters of representative task offloading schemes. A clear observation is that, despite previous efforts on scheme design for systems of *small* (less than 100 nodes) scales, there is little effort on the performance evaluation of *frequent* and *massive* task offloading operations, which are inevitable in fog computing based IoT systems.

An open, yet challenging question is, *how to evaluate the performance of such fog systems*? To be more specific, what is the expected processing time for a task, and how much resource will be consumed by the fog-based IoT system?

Considering that different offloading schemes can be adopted, and they differ in many aspects, as shown in Table 4.1, it is difficult, if not impossible, to measure the system performance by experiments. This difficulty can be further exacerbated by the large, and growing size of the fog-based IoT system, which is expected to cater for millions of IoT devices [1]. Consequently, the *scalability* of fog computing, as identified in the survey [40] on fog computing, is the crux of the performance evaluation problem. On one hand, the increase in network size permits more offloading options, which increases the offloading probability, and hence the

Table 4.1 Summary of representative task offloading literature in fog computing systems

Ref.	Approach	Scale	Mobility	Criteria	Communication	Task
Centralized schemes						
[10]	Optimization	< 100	Static	Delay/energy	BW=20 MHz,	Normal/Uniform/ Pareto on [2, 10]Mb
[6]	RL	< 10	Static	Delay	BW=2 MHz, Power=20 dBm	Fixed (500 Mb)
[28]	Matching Th.	< 10	Static	Hybrid	BW=10 MHz, variable power	Uniform on [300, 600]Kb
[5]	RL	< 10	Static	Delay/energy	Var. BW	Fixed (5 Kb)
Distributed schemes						
[18]	Game Th.	< 100	Static	Delay	Var. BW, Power=0.4W	Uniform on [0.1, 3.4]Mb
[27]	Deep RL	< 100	Mobile	Incentive	BW=10 MHz, fixed range	Uniform on [0.02, 0.2]Mb
Ours	Stochastic Th. analysis	$\leq 10^4$, Any(Th.)	Mobile	Delay/energy and hybrid	Multi-RAT, var. rate/range/power	Uniform on $\{1, 10, 10^2, 10^3\}$ Mb

BW Bandwidth, *RL* Reinforcement Learning, *Th.* Theory, *var.* variable

Quality-of-Service (QoS) of end applications. On the other hand, frequent data transmission as a result of higher offloading probabilities raises concerns in resource consumption. In addition, some participants in the fog, e.g., smart phones, may act as both task node and fog node, such that the total amount of data (to offload) also increases as the system becomes larger. Meanwhile, task nodes can have their own offloading criterion/algorithms, as a result of which the impact of network size is unclear.

Seemingly simple, large-scale performance evaluation is, nonetheless, challenging due to the sheer *scale* and *diversity* of the fog-based IoT system. On the application level, IoT applications are so broad and complex, that data are generated in various volume, form, and frequency [32]. On the network level, due to the open nature of fog computing, a task can be directed to a certain fog node by a variety of offloading criteria, e.g., centralized and cooperative [10], latency-minimizing [18, 36], energy-minimizing [35], latency-energy-tradeoff [19, 28], and so on. These two layers of complexity, i.e., openness and diversity, superpose, creating an unmanageable space for experiment evaluation in massive IoT systems [1]. Accordingly, simulation and theoretic analysis seem to be the only viable options to evaluate the performance of fog task offloading. However, both of these two approaches demand a *descriptive model* of the task offloading procedure, which can easily adapt to different offloading criteria, yet manageable to examine the scalability of the fog-based IoT system, to address the 'remedy or resource drain' concern.

4.1.3 Our Approach and Contributions

Instead of focusing on inter-node coordination, trade-offs, or attempting to reach a system-level optimum, which is both unnecessary and infeasible for such large and heterogeneous system, we address the performance evaluation problem from the perspective of data, and take a completely different approach, that is, finding out *how data blocks move* in the fog during the offloading process. The rationale behind this approach is that, resource consumption, in forms of spectrum, time, and storage space, are all driven by movements of data blocks, and can be examined through the traversing paths of tasks as data blocks. In regard to task offloading process in the fog, our contributions can be summarized as follows.

We define evaluation metrics, namely, *task lifetime*, *device effort*, and *network effort*, to quantify the end-to-end delay, the amount of consumed storage/processing resource owned by individual fog nodes, and communication resource shared by the edge network in a heterogeneous setting.

We propose a *gravity model* that views task/data blocks as particles, whose movements are driven by a *gravity force* between the task particle and candidate fog nodes, and influenced by the preference of task nodes. The gravity function can describe multiple, even hybrid offloading criteria for systems of variable scales, so performances of large-scale fog systems with heterogeneous criteria can be examined.

We find that, under the gravity model, which applies to distance-, energy-, and/or delay-oriented offloading, the lifetime and device effort of individual tasks decrease as $O(\frac{1}{N})$ over the network size N, while the network effort decreases much slower, even remain $O(1)$ when load-balancing is adopted. Both the analysis and numerical results indicate that, though the expected application delay improves as the system scales, the total consumed network resource may exceed what the system can provide, leading to a concern of resource drain.

4.2 System Model and Problem Formulation

In this section, we formally introduce the terminology and settings on nodes, tasks, and the offloading process, define the performance metrics from the perspective of data movements, and formulate the task offloading problem.

4.2.1 Network Model

Consider a fog system hosted by the 5G cellular network and wireless local area network (LAN). A miniature example is illustrated in Fig. 4.1. In this

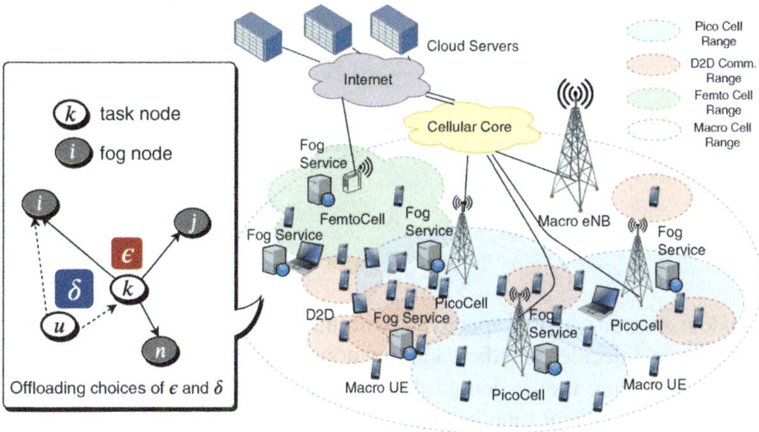

Fig. 4.1 An example of a fog system hosted by a heterogeneous wireless network, and a data offloading scenario (inner box on the left) in the fog

network,[1] the fog service is provided by BS's (of picocells and femtocells), AP's, and mobile devices (including smart phones, laptops, and so on), which we refer to as *fog nodes*.

4.2.1.1 Node Model

Any device that is authorized to offload tasks (to fog nodes) is referred to as a *task node*, so it is possible that a node (device) can act as a fog node and a task node at the same time. For instance, an end device k can choose to offload a heavy video trans-coding task ϵ to nearby the fog server deployed at the access point (AP) j, while accepting a file offloading (further to the cloud) task δ from a nearby Internet-of-Things (IoT) sensor u at the same time, as shown in the zoomed box on the left of Fig. 4.1.

Any device that is authorized to offload task (to fog nodes) is referred to as a *task node*. For example in Fig. 4.1, smart phone k can act as a fog node to task δ originated from task node u, e.g., uploading temperature measurements from smart home devices, and as the task node to generate task ϵ, e.g., video trans-coding.

Accordingly, each node $n \in E$ can act as two roles, as illustrated in Fig. 4.2: a fog node, who is associated with a CPU of processing capacity c_n, and a first-in-first-out (FIFO) queue; and/or a task node, who generates tasks at the rate of β. As a fog

[1] Considering resource bottlenecks mainly exist in wireless access networks, in this chapter, we restrict the scope of the fog paradigm to the network edge, despite the fact that wired connections between fog nodes and cloud servers are also viewed as part of the fog system in [17, 36].

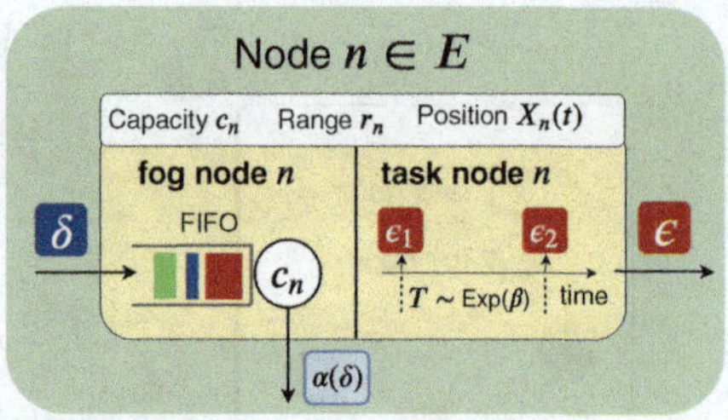

Fig. 4.2 Node model of fog-based task offloading system

node, node n announces[2] its capacity c_n within its transmission range r_n, while as a task node, node n decides whether, and to whom, a generated task will be offloaded. Since fog nodes may move, denote the position of node n as $X_n(t) : T \to \mathbb{R}^2$, where $t \in T$ is the current time instant.

4.2.1.2 Communication Model

For node $e \in E$, located at $X_e(t)$, denote its *neighbors* as

$$N_e(t) = \{i \in E \mid d(X_i(t), X_e(t)) \leq \min\{r_e, r_i\}\}, \tag{4.1}$$

where r_e is the transmission range of node e, and $d(X_e(t), X_n(t))$ measures the distance between node e and node i at time t. Suppose node e utilizes a wireless channel of bandwidth $\mathrm{BW}_{e\to n}(t)$ (Hz) to communicate with neighboring node $n \in N_e(t)$, located at $X_n(t)$, where the noise power measures at $P_0 = \mathrm{BW}_{e\to n} N_0 \mathrm{T}$, $N_0 = 1.38065 \times 10^{-23}$, and $\mathrm{T} = 290K$. Then if node e transmits with power $P_{e\to n}(t)$, and occupies a channel of width $\mathrm{BW}_{e\to n}(t)$, both of which remain the same throughout the offloading process, the transmission rate from node e to node n can be obtained as a function of three variables, i.e.,

$$R_{e\to n}(t) = \mathrm{BW}_{e\to n}(t) \log_2 \left(1 + \frac{P_{e\to n}(t) \cdot d\left(X_e(t), X_n(t)\right)^{-\kappa}}{\mathrm{BW}_{e\to n}(t) \cdot N_0 \cdot \mathrm{T}}\right), \tag{4.2}$$

where κ is the path-loss exponent.

Remark 4.1 The three variables in Eq. (4.2), that is, $\mathrm{BW}_{e\to n}(t)$, $P_{e\to n}(t)$ and $d\left(X_e(t), X_n(t)\right)$, are determined by the time instant t_ϵ, when task ϵ is generated. But their values can not be set freely. Instead, they need to comply with the RAT adopted in the current offloading process, and may change over different tasks if the IoT device is multi-mode, e.g., equipped with software-defined radio (SDR). Commonly used RAT's in a fog-based IoT system are listed in Table 4.2, where broadband and narrow-band RAT's are both presented.

[2] The processing capacity $c(n)$ (a few bytes long) may piggyback on link discovery messages, so task nodes can gather this information when it enters the communication range of node n.

Table 4.2 Radio access technologies (RAT) for IoT systems

RAT	Range	Channel bandwidth	Trans. power	Freq. bands	Peak rate
Broadband RAT					
5G-NR FR2 (FemtoCell)	<10 m	50, 100, 200, 400 MHz	10–20 dBm	mmWave (24–50 GHz)	20 Gbps
5G-NR FR2 (PicoCell)	<200 m	50, 100, 200, 400 MHz	20–24 dBm	mmWave (24–50 GHz)	20 Gbps
WiFi 6 (indoor-5GHz)	<50 m	20, 40, 80, 160 MHz	<20 dBm	ISM band (5 GHz)	10 Gbps
WiFi 6 (indoor-2.4GHz)	<50 m	20, 40 MHz	<20 dBm	ISM band (2.4 GHz)	2 Gbps
5G-NR FR2 (MicroCell)	<1 Km	50, 100, 200, 400 MHz	<23 dBm	Sub-6G (3.3–6 GHz)	1 Gbps
5G-NR FR1 (MicroCell)	<1 Km	5–100 MHz	<23 dBm	Sub-6G (3.3–6 GHz)	1 Gbps
LTE-M Cat2 (release 14)	<1 Km	5 MHz	20/23 dBm	LTE (0.9–2 GHz)	7 Mbps
Narrowband RAT					
NB-IoT (Cat NB2)	<1 Km	180 KHz	20/23 dBm	LTE (700–900 MHz)	200 Kbps
LoRa-WAN	<5 Km	125, 250 KHz	<14 dBm	ISM band (2.4, 5, 6 GHz)	50 Kbps

4.2.2 *Task Model*

Consider a task[3] $\epsilon \in \Omega$, which is a seven-element tuple $(t_\epsilon, L_\epsilon, \tau_\epsilon, S_\epsilon, D_\epsilon, \phi_\epsilon, \alpha_\epsilon)$, whose ranges and physical meanings can be found in Table 4.3. For example, the transmission of a file ϵ of size L_ϵ, to be offloaded from the source node $S_\epsilon = k$ to the destination node $D_\epsilon = n$, needs to be completed by time instant $t_\epsilon + \tau_\epsilon$. Task ϵ also requires a processing intensity ϕ_ϵ, and expects the execution result $\alpha(\epsilon)$ of volume $\alpha_\epsilon L_\epsilon$, where coefficient $\alpha_\epsilon \in (0, 1]$ is the ratio between the volume of data block $\alpha(\epsilon)$ and that of data block ϵ. For example, a video trans-coding task ϵ to compress a 1 minute 1080p video clip to a 360p version has volume $L_\epsilon = 165$ Mb, processing intensity $\phi_\epsilon \simeq 290$ cycles/bit [3, 20], and coefficient $\alpha_\epsilon = \frac{55 \text{ Mb}}{165 \text{ Mb}} \simeq 0.33$, as the processing result $\alpha(\epsilon)$, i.e., the 1 minute 360p video clip, is only 55 Mb in volume.

Among these attributes in Table 4.3, the first five are common attributes of data blocks in any dissemination process, while the last two capture the characteristics of data blocks specific to tasks. For example, a file ϵ of size L_ϵ, to be transferred from the source node $S_\epsilon = k$ to the destination node $D_\epsilon = n$, needs to be completed

[3] Technically, a task is a piece of computation work to be completed before a certain time instant, which usually contains a block of data to be processed. So we do not differentiate the two, and use 'task' and 'data block' interchangeably throughout this chapter.

Table 4.3 Attributes of a task ϵ (whose values are determined at time t_ϵ)

Attribute	Unit	Range	Physical meaning
t_ϵ	s	$\mathbb{R}^+$	The time instant that task ϵ is generated
L_ϵ	bits	$\mathbb{N}^+$	Volume of task ϵ (bits)
τ_ϵ	s	$\mathbb{R}^+$	Timeout/delay bound
S_ϵ	N/A	E	Task node who generated task ϵ
D_ϵ	N/A	E	Target fog node to offload
ϕ_ϵ	CPU cyles/bit	[100, 400]	Processing intensity [3, 20]
α_ϵ	N/A	(0, 1]	Coefficient of volume change after processing

by time instant $t_\epsilon + \tau_\epsilon$. As a task to be offloaded to node D_ϵ, data block ϵ has two more attributes in addition to the basic model, i.e., the processing intensity ϕ_ϵ and volume change coefficient $\alpha_\epsilon \in (0, 1]$, because a task ϵ needs to be *processed* for an expected result $\alpha(\epsilon)$, where function $\alpha : \{0, 1\}^{L_\epsilon} \to \{0, 1\}^{\alpha_\epsilon L_\epsilon}$ describes the processing procedure that maps task δ to the result $\alpha(\epsilon)$.

Three attributes of the task tuple, i.e., volume L_ϵ, processing intensity ϕ_ϵ, and volume change coefficient α_ϵ, may differ considerably among different types of tasks, e.g., a text-translation task has a small L_ϵ on the level of Kb, and coefficient $\alpha_\epsilon \simeq 1$, because the size of the translated file will be comparable with that of the original text file, both of which are small; virtual reality (VR)/augmented reality (AR) tasks have large L_ϵ, and coefficient $\alpha_\epsilon \simeq 1$, because the major processing object are images, which are relatively large in volume; while object recognition and cloud offloading tasks have medium L_ϵ, and small coefficient $\alpha_\epsilon \simeq 0$, because the expected result $\alpha(\epsilon)$ will be much shorter compared to the original data block ϵ.

All attributes of task ϵ are populated at their generation. Particularly, the offloading target D_ϵ decides the direction of ϵ's movements, and hence is most relevant to the delay and resource consumption performances. This key parameter is determined by the *offloading criterion*, accounting for the current status of the network, e.g., reachability, processing capacity, and cost of the fog service. For instance, under a delay-oriented criterion, task ϵ will most likely be offloaded to fog node i in Fig. 4.1, who can complete the task with the minimum delay, according to task node k's observation at t_ϵ. If task ϵ is viewed as a particle, it moves toward node i, as if drawn by a probabilistic *gravity force* between the two, and the resource consumed during the offloading process can be viewed as the *effort* along its moving path. This observation motivates us to develop effort-based performance metrics and the gravity-based offloading model, for heterogeneous communication options and offloading criteria, respectively.

4.2.3 *Performance Metrics through Data Movements*

The offloading process of task ϵ, is a course of transporting data block ϵ from its generator, task node k, to a fog node n, getting processed at n, and then returning

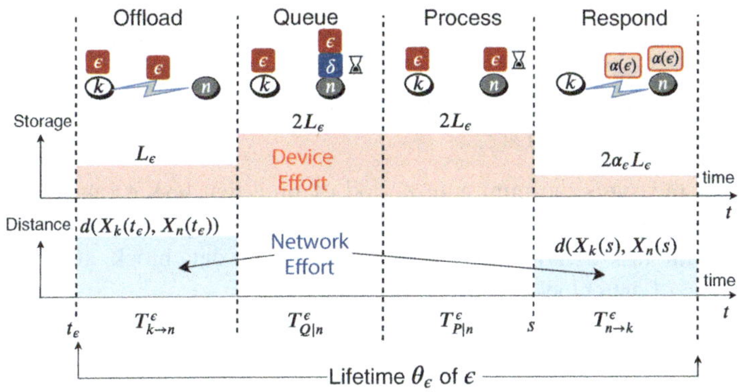

Fig. 4.3 Efforts spent to offload task ϵ from task node k to fog node n

the result $\alpha(\epsilon)$ back to its generator k. This process can be described by four stages: offloading (from k to n), queuing (at n), processing (at n), and responding (from n back to k), as shown at the top of Fig. 4.3. Note that the offloading target, i.e., fog node D_ϵ, is determined by the task node $S_\epsilon = k$ at t_ϵ, and may also take value $D_\epsilon = k$, which means k decides to process ϵ locally. In this case (local processing), the offloading process only contains the queuing and processing stages.

We consider performance[4] of task offloading processes from two aspects: the completion time of individual tasks, and the resource consumption during the offloading process. From the perspective of individual users, the generator S_ϵ of task ϵ needs to know the time that task ϵ is completed, i.e., the *lifetime* of task ϵ in the fog. From the perspective of the fog system, we define *device effort* and *network effort* to quantify the storage, and communication resource consumption, respectively.

Let $\mathcal{H}_\epsilon(t) \subset E$ denote the set of nodes that has data block ϵ locally stored at time t. Then with respect to ϵ, the state of each node $e \in E$ is a random variable $\epsilon_e(t) = \mathbb{1}_{\mathcal{H}_\epsilon(t)}(e) : \Omega \to \{0, 1\}$ in probability space $(\Omega, \mathcal{F}, \mathbb{P})$, and changes in $\mathcal{H}_\epsilon(t)$ can be represented by the random process $\{\boldsymbol{\epsilon}(t)\}_{t\in T}$ of random vectors $\boldsymbol{\epsilon}(t) = \{\epsilon_e(t)\}_{e\in E}$. For a generic data dissemination process, the movement trace of data block ϵ can be described by a process of random elements, $\{\mathcal{X}(t)\}_{t\in T}$, where $\mathcal{X}(t) = \bigcup_{e\in\mathcal{H}_\epsilon(t)}\{X_e(t)\}$, as a result of actions taken by the carrying nodes $\mathcal{H}_\epsilon(t)$. The movements of data block ϵ will stop when its *lifetime* θ_ϵ is reached.

Definition 4.1 Consider task ϵ generated by task node $S_\epsilon \in E$ at time t_ϵ. The **lifetime** of ϵ is define as the time between its generation, and the time that the last bit of its expected result $\alpha(\epsilon)$ reaches the task node S_ϵ, that is,

[4] Though task offloading in the fog paradigm is usually one-hop, i.e., direct transmission between the task node S_ϵ and a fog node D_ϵ, we define the performance metrics in a more general sense, such that they also apply to multi-hop offloading scenarios.

$$\theta_\epsilon := \min\Big\{\tau_\epsilon,\ \inf\{t \geq 0 \mid S_\epsilon \in \mathcal{H}_{\alpha(\epsilon)}(t)\} - t_\epsilon\Big\}, \tag{4.3}$$

where τ_ϵ is the timeout bound of task ϵ, and $\mathcal{H}_t(\alpha(\epsilon)) \subset E$ denotes the set of nodes that has $\alpha(\epsilon)$ locally stored at t.

Lifetime measures the minimum period of time that task ϵ (and its copies) is physically present in the fog system. It is upper bounded by the timeout value τ_ϵ (an attribute of the task determined at generation) of that data block, after which time the existence of data block ϵ is no longer necessary.

During its lifetime, three types of resources are consumed to complete the task offloading service, that is, communication resource (during the offloading and responding stage) shared by the edge network, storage resource (across the whole process) and processing resource (in the processing stage) located at individual devices. Inspired by the 'effort' analogy, we quantify the amount of resource consumption as the integral of the unit consumption ('power' of the gravity force) for a task over time. In this sense, no matter a task ϵ is offloaded or executed locally, the total amount of processing resource needed (unit: CPU cycles) equals to the product of data volume L_ϵ and processing intensity ϕ_ϵ, which is a constant value. Consumption of the storage and communication resource, on the other hand, are more difficult to analyze, as the unit consumption varies over time, for which the *device effort* and *network effort* are defined, respectively.

Remark 4.2 The concept of *effort* can be understood with an analogy in physics. If we view a data block ϵ as a solid object with mass $\phi_\epsilon L_\epsilon$, which is initially static at its generator S_ϵ. Then external forces have to be applied onto the data block to change its motion (from static to moving). Moreover, the motion of data block is accompanied with excessing work by the force, spent in the form of energy.

Definition 4.2 The **device effort** spent on task ϵ at time $t \in [t_\epsilon, t_\epsilon + \theta_\epsilon]$ is defined as the amount of local storage resource consumption (cumulative) in the carrying nodes,

$$\mathrm{DE}_\epsilon(t) := \int_{t_\epsilon}^{t} L_\epsilon |\mathcal{H}_\epsilon(s)| + \alpha_\epsilon L_\epsilon |\mathcal{H}_{\alpha(\epsilon)}(s)| \, \mathrm{d}s, \tag{4.4}$$

where $|\mathcal{H}_\epsilon(s)|$ is the number of copies of task ϵ stored on any device in the network. We then write $\mathrm{DE}_\epsilon := \mathrm{DE}_\epsilon(t_\epsilon + \theta_\epsilon)$ as the total device effort spent during the offloading process of task ϵ.

Note that $|\mathcal{H}_\epsilon(t)| = \|\epsilon(t)\|_1$ is a r.v. in probability space $(\Omega, \mathcal{F}, \mathbb{P})$, who takes range in $[0, |E|]$. The device effort $\mathrm{DE}_\epsilon(t)$ quantifies the amount of storage resource consumed by task ϵ up to time t. For instance, during the offloading process shown in Fig. 4.3, the amount of device effort spent on task ϵ is $2L_\epsilon$ during the queuing and processing stage, while it is L_ϵ during the offloading stage, and αL_ϵ during the responding stage. This is because both the task node k and the fog node n need to keep a copy of task ϵ to avoid transmission failure or processing failure, during the queuing and processing stages.

On the other hand, the *network effort* quantifies the amount of shared resource it takes the fog system to perpetuate the movements of data block ϵ over the air interface. By network resource, we mean communication resource of the provisioning wireless network, e.g., spectrum, resource blocks, transmission opportunities, and so on.

Definition 4.3 For task ϵ with lifetime θ_ϵ, the **network effort** of ϵ at time $t \in [t_\epsilon, t_\epsilon + \theta_\epsilon]$ is defined as the sum of every distance-transmission time product, that is,

$$\mathrm{NE}_\epsilon(t) := \sum_{s \in \mathcal{T}^\epsilon} \sum_{i \in \Delta\mathcal{H}_\epsilon(t)} T^\epsilon_{p(i)\to i} d\Big(X_i(s), X_{p(i)}(s)\Big), \tag{4.5}$$

where $\Delta\mathcal{H}_\epsilon(t) := \mathcal{H}_\epsilon(t) \setminus \mathcal{H}_\epsilon(t^-)$ denotes the set of new recipients of data block ϵ at time t, set $\mathcal{T}^\epsilon := \{s \in (t_\epsilon, t] \mid \Delta\mathcal{H}_\epsilon(t) \neq \phi\}$ denotes all the time instants when data block ϵ is transmitted over the air, and $p(i)$ denotes the node, from which i receives a copy of ϵ. We than write $\mathrm{NE}_\epsilon := \mathrm{NE}_\epsilon(t_\epsilon + \theta_\epsilon)$ as the total network effort spent during the entire offloading process of task ϵ.

We consider distance-time product rather than bandwidth-time product in the definition above, because spectrum resource (measured by channel bandwidth) may not be compared across systems of different RAT's, while the occupation of space and time resource is universal.

Remark 4.3 Device effort and network effort can not be added together directly, because they quantify consumption of resources by different types (and are hence in different units). Nonetheless, they jointly illustrate the amount of resources spent by the offloading process, and can be compared (separately) among different tasks.

In a more general sense, this formulation also applies to any data delivery process in a wireless network. The performance metrics are defined with respective to movements of data blocks, and hence easily apply to scenarios other than task offloading. In this chapter, we restrict ourselves to one-hop offloading of indivisible tasks, which are the most prevalent in a fog-based IoT system. To be more specific, we aim to find *how the expected task lifetime* $\mathbb{E}_\Omega(\theta)$, *total device effort* $\mathbb{E}(\sum_\epsilon \mathrm{DE}_\epsilon)$, *and network effort* $\mathbb{E}(\sum_\epsilon \mathrm{NE}_\epsilon)$ *scale with the system size* $N = |E|$, to address the 'remedy or resource drain' concern in the Fog-based IoT system.

4.3 How Data Move: The Gravity Model for Task Offloading

Seemingly straight-forward, the three performance metrics are in fact quite challenging to evaluate for a task offloading process, due to the following reasons.

- *Crux of diversity*: A task node may employ different RAT's, as well as different offloading criteria with or without inter-nodal coordination, resulting in an unmanageable space for performance evaluation.
- *Curse of openness*: The lifetime, device effort and network effort of an individual task are inevitably influenced by other tasks' offloading processes, which are unclear as the size of the system N grows.

The first challenge on diversity calls for a generic offloading model, while the second on openness requires meticulous analysis on the competition among tasks. In this section, we introduce a *gravity*-based offloading model to tackle the crux of diversity, and leave analysis to the next section. Under the gravity setting, different offloading criteria can be written as gravity functions quantifying the 'suitableness' of candidate fog nodes. Then, each task node selects a qualified candidate with a *probability* that is proportionate to the gravity force.

4.3.1 An Offloading Procedure Under Gravity Rule

The state transitions of task ϵ during its offloading process are shown in Fig. 4.4. And the offloading process, as shown in Fig. 4.3, can be described as follows:

1. At time t_ϵ, node $S_\epsilon = k \in E$ generates task ϵ, so data block ϵ is only stored by nodes in set $\mathcal{H}_\epsilon(t_\epsilon) = \{k\}$.
2. Task node k chooses the offloading target $D_\epsilon = n$ out of the candidate set $N_k(t_\epsilon) \cup \{k\}$, with the *offloading probability* $\mathbb{P}(D_\epsilon = n)$.
3. Task ϵ is then transmitted from node k to node n at data rate $R_{k\to n}(t_\epsilon)$. At this point, data block ϵ is stored by nodes in $\mathcal{H}_\epsilon(t_\epsilon + T^\epsilon_{k\to n}) = \{n, k\}$. This step is omitted if k decides to execute ϵ locally, and $T^\epsilon_{k\to n} = 0$.
4. Upon the reception of task ϵ, node n puts ϵ in a queue if its processor is busy, and starts executing ϵ when all of the tasks queued before are completed.
5. When task ϵ is completed (after queuing delay $T^\epsilon_{Q|n}$ and processing delay $T^\epsilon_{P|n}$), the corresponding result $\alpha(\epsilon)$ of size $\alpha_\epsilon L_\epsilon$ will be transmitted back to its corresponding task node k at the data rate of $R_{n\to k}(t)$.
6. At time instant $t_\epsilon + T^\epsilon_{k\to n} + T^\epsilon_{Q|n} + T^\epsilon_{P|n} + T^\epsilon_{n\to k}$, the task response $\alpha(\epsilon)$ is returned to node k at the same data rate $R_{k\to n}(t_\epsilon)$.

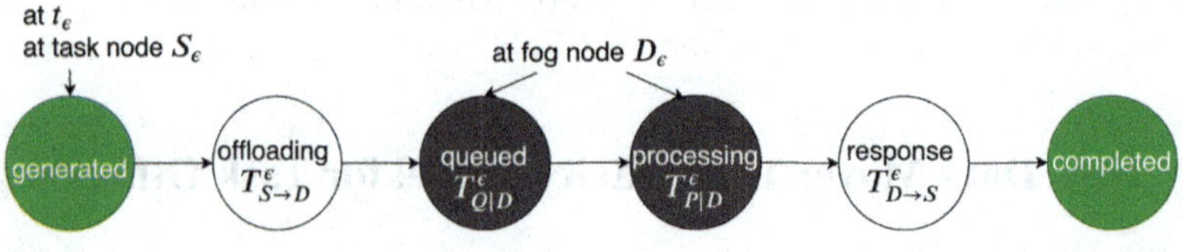

Fig. 4.4 An offloading process from the perspective of task ϵ. Note that S_ϵ can be the same as D_ϵ for local processing, in which case transmission time $T^\epsilon_{S\to D} = T^\epsilon_{D\to S} = 0$

The first feature of the gravity rule model is allowing probabilistic target selection for load-balancing purposes, which is captured by the offloading probability.

Definition 4.4 The **offloading (to n) probability** $\mathbb{P}(D_\epsilon = n)$, that is, probability that fog node $n \in E$ is chosen by task node S_ϵ as the offloading target for task ϵ, is defined as

$$\mathbb{P}(D_\epsilon = n) := \begin{cases} 0, & \text{if } \sum_{i \in N_k(t_\epsilon)} F(\epsilon, i) = 0, \\ \frac{\rho(F(\epsilon,n))}{\sum_{i \in N_k(t_\epsilon)} \rho(F(\epsilon,i))}, & \text{otherwise,} \end{cases} \tag{4.6}$$

where *gravity* $F : \Omega \times E \to [0, \infty)$ is a non-negative function, and the selectiveness $\rho : [0, \infty) \to [0, \infty)$ is an increasing, convex function, both chosen by task node S_ϵ.

The gravity function F reflects task node S_ϵ's unique objective in task offloading, e.g., energy saving, latency reducing, and so on. Any candidate offloading target must impose a positive gravity on the task, which jointly considers the communication range and suitableness of the fog node, as shown in Fig. 4.5. Note that when none of the candidate fog nodes imposes a positive gravity (e.g., the transmission time alone is longer than the timeout value τ_ϵ) on task ϵ, i.e., the first case in Eq. (4.6), task ϵ will be processed locally, i.e., $D_\epsilon = S_\epsilon = k$.

The selectiveness function $\rho(\cdot)$ allows the task node to fine-tune its preferences, that is, how reluctant it is against adjustments due to load balancing reasons. It may take any form, as long as it is an increasing function. For example, we may set $\rho(x) = x^\gamma$, where γ is the selectiveness coefficient. The larger γ, the more likely task node S_ϵ will favor fog nodes imposing the highest gravity, i.e., it is the more reluctant to settle for 'sub-optimal' offloading choices. Together, these two functions $\{F, \rho\}$ compose an offloading profile of task node S_ϵ, in which different offloading criteria can be described in a unified form to address the crux of diversity.

From the perspective of probabilistic offloading setting, our model bears some resemblance with [18], who proved the existence of an equilibrium, and solved

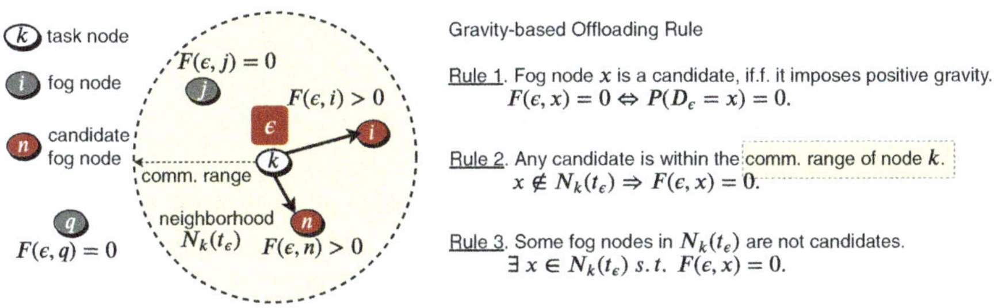

Fig. 4.5 An example of the gravity-based offloading rule: every fog node $i \in N_k(t_\epsilon) \cup \{k\}$ (including the task node k itself) imposes a non-negative gravity force $F(\epsilon, i)$ on task ϵ, and the offloading probability from k to i is proportionate to the gravity force $F(\epsilon, i)$ defined in Eq. (4.6)

the optimization problem of delay minimization with a distributed algorithm. Nonetheless, our model serves a different purpose, that is, describing heterogeneous offloading criteria, and is hence more general, as it is not restricted to delay-oriented offloading as in [18]. Moreover, the 'optimal' and 'sub-optimal' are determined solely by the task node of interest, without any collaboration that may induce overhead, nor information of other tasks. This setting of local decision is especially practical for IoT devices that lack the capacity for cooperation, e.g., due to power constraints. Next, we discuss the form of gravity functions for typical and generic offloading schemes.

The probabilistic setting addresses the identified challenges in modeling the offloading process from the following aspects:

1. *Diverse offloading criteria.* The definition of offloading probability provides two options with respect to modeling different criteria. First, gravity function $F(\epsilon, i)$ can be set by the task node to adapt to its unique requirements. Second, selectiveness function $\rho(\cdot)$ allows the task node to fine-tune how selective it is, or how reluctant it is to adjustments due to load-balancing.
2. *Competition and adjustment.* The probabilistic setting describes the offloading choices due to competition among tasks, and achieves 'automatic' load-balancing among fog nodes. Under this setting, tasks are not always offloaded to the seemingly 'optimal' choice (fog node imposing the largest gravity force) from the (possibly incomplete) view of the task node, to avoid a small number of powerful fog nodes being overloaded by excessive task offloading.

4.3.2 Typical and Generic Gravity Rules

The gravity force $F(\epsilon, n)$ is meant to quantify the 'attractiveness' of fog node n, in regard to the processing of task ϵ, which can be freely defined by any task node. Considering most offloading criteria have an optimization goal, such as delay, energy consumption, etc., we define the gravity function as the performance gain achieved by offloading compared to local processing, or the remaining allowance against a performance budget. We list a few example rules for typical scenarios.

4.3.2.1 Distance-Oriented Gravity Rule

The simplest offloading scheme is to always offload to the nearest fog node, as long as there is one in the communication range of the task node, regardless of status of the fog nodes. This simple rule applies to offloading processes from IoT devices without local processing power, in which energy spent on data transmission over the air interface is the major concern, and hence the preference on short distances. The gravity force for distance-based offloading can be written as

$$F(\epsilon, n) = \frac{1}{d\left(X_k(t_\epsilon), X_n(t_\epsilon)\right)}, \quad \forall n \in N_e(t_\epsilon). \tag{4.7}$$

To always select the nearest fog node, the probability mapping function $\rho(x)$ can be set to $\rho(x) = x^\gamma$ with a large γ, because the larger the selectiveness coefficient γ is, the more likely the task node chooses the nearest fog node.

4.3.2.2 Delay-Oriented Gravity Rule

As shown in the state transition diagram (Fig. 4.4), for a task ϵ generated by $S_\epsilon = k$, three out of the four delay components can be determined, once the offloading decision $D_\epsilon = n$ is made at time t_ϵ: the uplink transmission delay $T^\epsilon_{S\to D}$, the processing delay $T^\epsilon_{P|D}$, and the downlink transmission delay $T^\epsilon_{D\to S}$.

The queuing delay $T^\epsilon_{Q|n}$, on the other hand, also depends on the offloading processes of *other* tasks. It is bounded by

$$0 \le T^\epsilon_{Q|n} \le \frac{Q_n(t_\epsilon + T^\epsilon_{k\to n}) + Y_n}{c_n} < \frac{Q_n(t_\epsilon) + Y_n}{c_n}, \tag{4.8}$$

where $Q_n(t)$ denotes the processing queue length of fog node n at time t, and r.v. $Y_n \in [0, \infty)$ is the workload offloaded to node n, during the transmission of task ϵ, i.e., interval $[t_\epsilon, t_\epsilon + T^\epsilon_{k\to n}]$. Note that task node k, as a fog node, is also accepting tasks offloaded from other task nodes, so local execution may induce queuing delay as well.

Then, the corresponding gravity function can be defined as

$$\begin{aligned} F(\epsilon, n) &= \left[\tau_\epsilon - \left(T^\epsilon_{k\to n} + T^\epsilon_{P|n} + T^\epsilon_{n\to k}\right)\right]^+ \\ &= \left[\tau_\epsilon - \left(\frac{(1+\alpha_\epsilon)L_\epsilon}{R_{k\to n}(t_\epsilon)} + \frac{L_\epsilon \phi_\epsilon}{c_n}\right)\right]^+, \end{aligned} \tag{4.9}$$

where c_n (cycles/s) is the capacity of candidate fog node n, and $[x]^+ = \max\{x, 0\}$.

This rule applies to many IoT applications, as reducing application latency is the primary motivation [31] of fog computing. In this case, devices tend to transmit with the highest data rate that a RAT provides, so the number of variables in Eq. (4.2) can be further reduced, that is,

$$R_{k\to n}(t) = g_d\left(d\left(X_n(t), X_k(t)\right)\right). \tag{4.10}$$

The reason behind this reduction is that the possible value of $P_{k\to n}(t_\epsilon)$ and $\mathrm{BW}_{e\to n}(t_\epsilon)$ are actually determined by the range of a RAT, as indicated by Table 4.2. For instance, RAT 5G-NR femto-cell and pico-cell, although reaching a staggering 5–40 Gbps in lab environments [9, 16] with mmWave technology, only covers tens of meters of communication, due to the acute attenuation in the 26–60 GHz bands. For the range of hundreds of meters, WiFi and LTE/5G micro-cell are available, offering a transmission rate at ten to a hundred Mbps [14]. Finally, for the range of over one Kilometer, LTE-M (4–7 Mbps) and narrow band RAT's can be adopted, such as NB-IoT (200 Kbps) and LoRA WAN (50 Kbps). In this sense, $g(\cdot)$ is a piece-wise function over the distance, and we can further simplify Eq. (4.10) as $g_d(d\,(X_n(t), X_k(t))) \triangleq R_{max} - \psi d(X_n(t), X_k(t))$ in the analysis, where R_{max} is the maximum transmission rate that can be achieved in the fog system, and ψ (Kbps/m) is the rate attenuation factor.

4.3.2.3 Energy-Oriented Gravity Rule

For pocketable, battery-operated task nodes, e.g., IoT sensors, energy consumption is a major concern. So the goal of offloading is to conserve as much energy as possible, i.e., task ϵ may be offloaded to node n only if the energy spent on transmission $P_{k\to n}T^{\epsilon}_{k\to n}$ is less than the energy consumed by local execution $E^{\epsilon}_{k} = \xi\phi_\epsilon L_\epsilon$, where constant ξ (Joule/cyle) is the amount of energy consumed by each CPU cycle. Then the gravity function can be defined as the amount of energy saved if the corresponding task is offloaded, that is,

$$
\begin{aligned}
F(\epsilon, n) &= \left[\xi\phi_\epsilon L_\epsilon - \frac{P_{k\to n}(t_\epsilon)L_\epsilon}{R_{k\to n}(t_\epsilon)}\right]^+ \\
&= \left[\xi\phi_\epsilon L_\epsilon - \frac{L_\epsilon}{g_e\left(\frac{P_{k\to n}(t_\epsilon)}{\mathrm{BW}_{k\to n}(t_\epsilon)}; d\,(X_k(t_\epsilon), X_n(t_\epsilon))\right)}\right]^+,
\end{aligned}
\tag{4.11}
$$

where duel-variate function $g_e(x; y) = \frac{1}{x}\log_2\left(1 + \frac{xy^{-\kappa}}{N_0\mathrm{T}}\right)$.

Note that $g_e\left(\frac{P_{k\to n}(t_\epsilon)}{\mathrm{BW}_{k\to n}(t_\epsilon)}; d\,(X_k(t_\epsilon), X_n(t_\epsilon))\right) = \frac{P_{k\to n}(t_\epsilon)}{R_{k\to n}(t_\epsilon)}$ in value. It is different from the delay-oriented rule, where the single-variate function $g_d(d\,(X_k(t_\epsilon), X_n(t_\epsilon))) = R_{k\to n}(t_\epsilon)$. But they are both functions of the distance.

4.3.2.4 Generic Form of the Gravity Function

For the ease of analysis, we generalize the aforementioned gravity functions into the following unified form:

$$
F(\epsilon, n) := f\left(A_\epsilon - \frac{B_\epsilon}{c_n} + \frac{G_\epsilon}{g\,(d(X_k(t_\epsilon), X_n(t_\epsilon)))}\right), \tag{4.12}
$$

quantities in which have the following properties:

- Function $f : \mathbb{R} \rightarrow \mathbb{R}$ is non-decreasing; $f(a) > 0$ for any $a > 0$; and $f(a) = 0$ otherwise. The simplest form of f is $f(x) = [x]^+$.
- Function $g : \mathbb{R}^+ \rightarrow \mathbb{R}$ is a non-decreasing function of the transmission distance $d(X_k(t_\epsilon), X_n(t_\epsilon))$.
- Quantities A_ϵ, B_ϵ, and G_ϵ are non-negative constants, which are determined solely by the attributes of task ϵ, and satisfy constraint $B_\epsilon + G_\epsilon > 0$.

The rationale of Eq. (4.12) is that, despite their differences, the gist of all offloading rules is to find a fog node with the highest capacity, the shortest processing delay, and the least communication cost in terms of energy (i.e., that is located near the task node's current position). Together with the selectiveness function $\rho(\cdot)$, this generic gravity form in Eq. (4.12) can describe all the aforementioned gravity rules, indicating the versatility of the proposed model. Particularly, for the distance-oriented rule in Eq. (4.7), $A_\epsilon = B_\epsilon = 0$, $G_\epsilon = 1$, and function $g(x) = x$, while for the delay-oriented rule in Eq. (4.9), $A_\epsilon = \tau_\epsilon$, $B_\epsilon = \phi_\epsilon L_\epsilon$, $G_\epsilon = (1 + \alpha_\epsilon)L_\epsilon$, and $g(x) = -g_d(x)$, whose absolute value $|g(x)|$ measures the data transmission rate at distance x. For energy-oriented offloading, function $g(\cdot) = -g_e(\cdot;\cdot)$ is multivariate, in which $\frac{P_{k\rightarrow n}(t_\epsilon)}{\mathrm{BW}_{k\rightarrow n}(t_\epsilon)}$ is determined by the power control mechanism.

Remark 4.4 Note in Eqs. (4.7)–(4.9), the physical meanings of gravity force $F(\epsilon, n)$ are different, and measured in different units. Nonetheless, any joint scheme of the distance-oriented and delay-oriented rules can be described as the linear combination of Eqs. (4.7) and (4.9), and it still has the same form as Eq. (4.12).

4.3.3 Offloading Probability Under the Generic Gravity Rule

Fog systems are expected to include a large number of devices, so it is highly probable that multiple offloading candidates co-exist for a task in the vicinity of the task node. Consequently, the actual efforts of the offloading process depend on *whether to offload*, and *if so, which fog node to choose*. To address this issue, we first discuss the offloading options and their probabilities for a given task ϵ.

Consider $\epsilon = (t_\epsilon, L_\epsilon, \tau_\epsilon, S_\epsilon, D_\epsilon, \phi_\epsilon, \alpha_\epsilon)$, i.e., a task[5] of size L_ϵ, generated at t_ϵ by $S_\epsilon = k$, and with processing intensity ϕ_ϵ and volume change coefficient α_ϵ. For any fog node $n \in E$, let r.v. $V_n = d(X_k(t_\epsilon), X_n(t_\epsilon))$ denote the distance between node n and node k, with probability density function (PDF) $f_V(v)$ $(v \in [0, +\infty))$, and r.v. C_n denote its processing capacity, with PDF $f_C(c)$ $(c \in [c_{min}, c_{max}])$. First, we present the probability for local processing, that is, task ϵ will not be offloaded to any candidates in node k's neighborhood.

Lemma 4.1 *Under the generic gravity function defined in Eq.* (4.12), *the probability that task node $S_\epsilon = k$ decides to process task ϵ locally is given by*

[5] For simplicity reasons, we assume the timeout value $\tau_\epsilon = \frac{L_\epsilon \phi_\epsilon}{c_{S_\epsilon}}$.

$$\mathbb{P}(D_\epsilon = k) = 1 - \left(1 - \mathbb{P}^\epsilon_{pos}\right)^{N-1}. \tag{4.13}$$

Particularly, when $A_\epsilon \neq 0$, quantity $\mathbb{P}^\epsilon_{pos}$ can be calculated as

$$\mathbb{P}^\epsilon_{pos} = \int_{\frac{B_\epsilon}{A_\epsilon}}^{c_{max}} \int_0^{g^{-1}\left(\frac{cG_\epsilon}{cA_\epsilon - B_\epsilon}\right)} f_V(v)\mathrm{d}v f_C(c)\mathrm{d}c, \tag{4.14}$$

where c_{max} is the maximum capacity among fog nodes in E.

Proof Quantity $\mathbb{P}^\epsilon_{pos}$ is the probability that a randomly chosen fog node n from set $E \setminus \{k\}$ imposes a strictly positive gravity force on task ϵ, which explains the form of Eq. (4.13). For distance-oriented offloading, any fog node will impose a positive gravity force on ϵ, so we focus on the latter two cases, i.e., $A_\epsilon \neq 0$. For any fog node $n \in E \setminus \{k\}$ to be considered as an offloading candidate, two conditions need to hold simultaneously: (1), node n resides in k's communication range, which means $g(V_n) < 0$; (2), gravity force $F(\epsilon, n) > 0$, which means there must be a performance gain if task ϵ is offloaded to fog node n. These translate to the following sufficient conditions:

$$\begin{cases} C_n > \frac{B_\epsilon}{A_\epsilon}, & \text{Condition (a)}, \\ \frac{G_\epsilon}{g(V_n)} > \frac{B_\epsilon}{C_n} - A_\epsilon, & \text{Condition (b)}. \end{cases} \tag{4.15}$$

From Eq. (4.15), each capacity value $c \in [c_{min}, c_{max}]$ corresponds to a threshold distance $v_{max}(c) = g^{-1}\left(\frac{cG_\epsilon}{B_\epsilon - cA_\epsilon}\right)$, beyond which Condition (b) will no longer hold, hence the four integral limits in Eq. (4.14). Note that the threshold distance v_{max} is smaller than or equal to the communication range, because when Condition (a) in Eq. (4.15) holds, the right-hand-side of Condition (b) is guaranteed to be negative, such that the requirement on communication range $g(V_n) < 0$ is satisfied automatically. Then, because the distribution of distance V_n and capacity C_n are independent of each other, probability $\mathbb{P}_{pos}$ can be calculated as Eq. (4.14). □

Lemma 4.1 provides a criterion of selecting offloading candidates from $N_k(t_\epsilon)$ for task ϵ, as well as the probability that ϵ will be processed locally at its generator k. It is a generic result, in the sense that for different scenarios, e.g., pedestrians walking in a region under a random walk model, or vehicles moving along a highway, this lemma gives numeric results, as long as the distributions of node distance V_n and processing capacity C_n are known. This is especially true when node statistics (captured by the distributions of distance V_n and processing capacity C_n) are known. Compared with task statistics [18], which are highly diverse and prone to privacy leak, node statistics are much easier to obtain. With node statistics, we discuss the offloading probability $\mathbb{P}(D_\epsilon = n)$ defined in Eq. (4.6). It is also worth noticing that the sum of offloading probabilities (equivalently $1 - \mathbb{P}(D_\epsilon = k)$) can be obtained with statistics of the node ensemble, instead of exact parameters of each individual node. Consequently, it is possible for a network operator to estimate the offloading performance through analysis by the proposed model, instead of lengthy simulation or costly experiments.

Next, we discuss the offloading probability to a specific fog node n, whose capacity c_n and distance $d(X_n(t_\epsilon), X_k(t_\epsilon)) \stackrel{\Delta}{=} v_n$ are known by task node k at time t_ϵ.

Theorem 4.1 *Let $\mu_C = \int_{c_{min}}^{c_{max}} c f_C(c) dc$ denote the mean capacity of fog nodes in E, and $F_V(v) = \int_0^v f_V(s) ds$ denote the CDF of distance V between task node k and a randomly chosen fog node at time t_ϵ. The probability that task ϵ will be offloaded to a fog node n with capacity c_n and distance v_n, i.e., $\mathbb{P}(D_\epsilon = n)$, is upper bounded by*

$$\mathbb{P}(D_\epsilon = n) \le \frac{\rho\left(F(\epsilon, n)\right)}{\mathrm{LB}(v_{max})\rho\left(A_\epsilon - \frac{B_\epsilon}{\mu_C} + \frac{G_\epsilon}{F_V(v_{max})}\int_0^{v_{max}} \frac{f_V(v)}{g(v)} dv\right)}, \tag{4.16}$$

where $v_{max} = g^{-1}(\frac{c_{max} G_\epsilon}{B_\epsilon - c_{max} A_\epsilon})$ is the maximum possible distance between a candidate node and task ϵ at time t_ϵ, and $\mathrm{LB}(v_{max}) = N F_V(v_{max}) f_C(c_{max}) \frac{-B_\epsilon G_\epsilon g(v_{max})}{(A_\epsilon g(v_{max}) + G_\epsilon)^2}$ is an lower bound of the total number of offloading candidates.

Proof The key to derive the offloading (to n) probability $\mathbb{P}(D_\epsilon = n)$ in Eq. (4.6), is finding the denominator $\sum_{i \in N_k(t_\epsilon)} \rho\left(F(\epsilon, i)\right)$ in the definition of the offloading probability.

First, as $\rho(\cdot)$ is a convex function, e.g., $\rho(x) = x^\beta$, we have

$$\sum_{i \in N_k(t_\epsilon)} \rho\left(F(\epsilon, i)\right) \stackrel{\text{Jensen}}{\ge} |N_+(t_\epsilon)| \rho\left(\frac{\sum_{i \in N_+(t_\epsilon)} F(\epsilon, i)}{|N_+(t_\epsilon)|}\right), \tag{4.17}$$

where $N_+(t_\epsilon) \subset N_k(t_\epsilon)$ is the set of candidate fog nodes that impose a positive gravity force on task ϵ.

The next step is to obtain $|N_+(t_\epsilon)|$ and $\sum_{i \in N_k(t_\epsilon)} F(\epsilon, i)$, that is, the total number of candidate nodes, and the total gravity force imposed by all candidates on task ϵ.

Observing that the gravity force only concerns the capacity C_i and distance V_i of a node i, and both quantities are additive, we resort to a divide-and-conquer method for obtaining $|N_+(t_\epsilon)|$ and $\sum_{i \in N_+(t_\epsilon)} F(\epsilon, i)$, illustrated in Fig. 4.6.

To be more specific, we divide the disk area (beige shades) centered at $X_k(t_\epsilon)$ into circular regions of width Δv, as shown in Fig. 4.6. Consider the circular region with inner radius $v \in [0, v_{\max}]$, as indicated by the blue ring with width Δv. When Δv is small, the probability that a node falls into this region can be approximated as $f_V(v)\Delta v$. The probability that an arbitrary fog node i is located inside this circular region, and satisfies the criterion in (4.15) is then

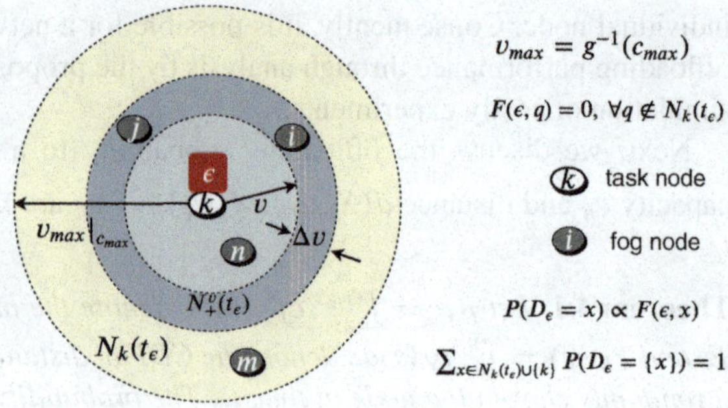

Fig. 4.6 Proof technique in Theorem 4.1: dividing the neighborhood of task node k into multiple rings of width Δv

$$p_0(v) = f_V(v)\Delta v \int_{\frac{B_\epsilon g(v)}{A_\epsilon g(v)+G_\epsilon}}^{c_{max}} f_C(c)\mathrm{d}c = f_V(v)\Delta v \left[1 - F_C\left(\frac{B_\epsilon g(v)}{A_\epsilon g(v) + G_\epsilon}\right)\right], \tag{4.18}$$

where $F_C(c)$ is the CDF of r.v. C that represents the capacity of a randomly chosen fog node from set E.

Denote the set of candidate fog nodes inside the blue ring as $N_+^v(t_\epsilon)$. It is easy to see that r.v. $|N_+^v(t_\epsilon)| \sim \mathbf{B}(N, p_0(v))$, with mean

$$\mathbb{E}(|N_+^v(t_\epsilon)|) = N f_V(v)\Delta v \left[1 - F_C\left(\frac{B_\epsilon g(v)}{A_\epsilon g(v) + G_\epsilon}\right)\right]. \tag{4.19}$$

Then the expected total number of candidates inside the beige-shaded disk, $\mathbb{E}(|N_+(t_\epsilon)|)$, can be lower bounded as

$$\begin{aligned}\mathbb{E}(|N_+(t_\epsilon)|) &= \mathbb{E}\left(\int_0^{v_{max}} |N_+^v(t_\epsilon)|\mathrm{d}v\right) \overset{\text{Fubini}}{=} \int_0^{v_{max}} \mathbb{E}\left(|N_+^v(t_\epsilon)|\right)\mathrm{d}v \\ &\geq N F_V(v_{max}) f_C(c_{max}) \frac{-B_\epsilon G_\epsilon g(v_{max})}{(A_\epsilon g(v_{max}) + G_\epsilon)^2},\end{aligned} \tag{4.20}$$

where $g(v_{max}) < 0$ is guaranteed by the candidate criterion in Eq. (4.15), such that $\mathbb{E}(|N_+(t_\epsilon)|)$ is always positive. Meanwhile, the expected number of candidates $\mathbb{E}(|N_+(t_\epsilon)|)$ is also upper-bounded by the total number of nodes within the beige disk, which also contains nodes that fail criterion Eq. (4.15), that is,

$$\mathbb{E}(|N_+(t_\epsilon)|) \leq N F_V(v_{max}). \tag{4.21}$$

On the other hand, the total gravity force on task ϵ can be calculated as the sum of all Δv rings within the disk of radius v_{max}, that is,

$$\mathbb{E}\left(\sum_{i\in N_k(t_\epsilon)} F(\epsilon,i)\right) \overset{\text{Fubini}}{=} \int_0^{v_{max}} \mathbb{E}\left(\sum_{i\in N_k^v(t_\epsilon)} F(\epsilon,i)\right) \mathrm{d}v$$

$$\geq \int_0^{v_{max}} N f_V(v)\left[1 - F_C\left(\frac{B_\epsilon g(v)}{A_\epsilon g(v)+G_\epsilon}\right)\right] \mathbb{E}\left(F(\epsilon,i)|V_i = v\right) \mathrm{d}v$$

$$\geq N\int_0^{v_{max}} \left[1 - F_C\left(\frac{B_\epsilon g(v)}{A_\epsilon g(v)+G_\epsilon}\right)\right]\left(A_\epsilon + \frac{G_\epsilon}{g(v)} - \frac{B_\epsilon}{\mu_C}\right) \mathrm{d}F_V(v)$$

$$\geq N\left(A_\epsilon - \frac{B_\epsilon}{\mu_C}\right) F_V(v_{max}) + NG_\epsilon \int_0^{v_{max}} \frac{f_V(v)}{g(v)}\mathrm{d}v, \tag{4.22}$$

where $\mu_C = \mathbb{E}(C)$ is the mean capacity of fog nodes in set E.

Consequently, the probability that task ϵ is offloaded to node n is then upper bounded by

$$\mathbb{P}(D_\epsilon = n) \leq \frac{\rho\left(F(\epsilon,n)\right)}{|N_+(t_\epsilon)|\rho\left[\frac{\mathbb{E}\left(\sum_{i\in N_+(t_\epsilon)} F(\epsilon,i)\right)}{|N_+(t_\epsilon)|}\right]}. \tag{4.23}$$

Plugging Eq. (4.22) on the sum of the gravity, upper bound Eq. (4.21), and lower bound Eq. (4.20) of the total number of candidate fog nodes into Eq. (4.23) yields the upper bound in Eq. (4.16). □

We highlight two implications from the upper bound in Theorem 4.1:

1. *Local action and global status.* With this theorem, we establish a link between a local action (task node k chooses fog node n to offload task ϵ) to the global status of the system, because in this process, both device effort and network effort for task ϵ take changes, which in turn affects the offloading options, and hence the efforts spent for other tasks. In other words, with the set of probabilities $\{\mathbb{P}(D_\epsilon = n)\}_{n\in E}$, we are now able to examine interactions of multiple tasks, and their movements after time instant t_ϵ in the fog system.
2. *Invariant property.* Regardless of the form of the gravity function and the selectiveness function, offloading probability $\mathbb{P}(D_\epsilon = n) = O(\frac{1}{N})$. In other words, this probability is asymptotically bounded above by $\frac{1}{N}$, when the selectiveness function $\rho(\cdot)$ takes a polynomial form, i.e., $\rho(x) = x^\gamma + x^{\gamma-1} + \cdots + x + 1$.

4.4 Bounds and Scaling Laws of Performances

Recall that, in the one-hop data offloading process, the lifetime, device effort and network effort of a single task ϵ can be obtained by their definitions once the offloading target D_ϵ is chosen to be fog node $n \in E$, that is,

$$\begin{cases} \theta_\epsilon^n & = T_{k\to n}^\epsilon + T_{Q|n}^\epsilon + T_{P|n}^\epsilon + T_{n\to k}^\epsilon, \\ \mathrm{DE}_\epsilon^n(\theta_\epsilon) & = L_\epsilon \left(T_{k\to n}^\epsilon + 2T_{Q|n}^\epsilon + 2T_{P|n}^\epsilon \right) + \alpha_\epsilon L_\epsilon T_{n\to k}^\epsilon, \\ \mathrm{NE}_\epsilon^n(\theta_\epsilon) & = T_{k\to n}^\epsilon d(X_k(t_\epsilon), X_n(t_\epsilon)) + T_{n\to k}^\epsilon d(X_k(\theta_\epsilon + t_\epsilon), X_n(\theta_\epsilon + t_\epsilon)), \end{cases} \tag{4.24}$$

as illustrated by the example in Fig. 4.3. These equations permit us to analyze the expected lifetime and efforts for offloading a single task ϵ through Theorem 4.1.

The three performance metrics on the offloading process of a task ϵ can only be evaluated, when fog node n is chosen as the offloading target D_ϵ. In this case, every term in Eq. (4.24) is only dependent on parameters (c_n, V_n) of fog node n, and the task ϵ, except for the queuing delay $T_{Q|n}^\epsilon$, which is the result of multiple offloading processes of tasks other than task ϵ.

4.4.1 Expected Queuing Delay $\mathbb{E}(T_{Q|n}^\epsilon)$

Consider another task δ that is also offloaded to fog node n, which takes place with probability $\mathbb{P}(D_\delta = n)$. Its processing may add to the queuing delay of task ϵ, if and only if the last bit of δ arrived at node n before the last bit of ϵ, such that δ will be processed before ϵ, that is, $t_\epsilon < t_\delta + T_{S_\delta\to n}^\delta < t_\epsilon + T_{S_\epsilon\to n}^\epsilon$.

According to the task node model described in Sect. 4.2, the data traffic generation process of every node in set E is stationary, i.e., the distribution of intervals between two consecutive tasks generated by the same node does not change over time, and the mean interval length is $\frac{1}{\beta}$. In other words, the system will generate $N\beta$ tasks on average, creating $N\beta\mu_L\mu_\phi$ processing load, for each time step. If the total processing capacity $N\mu_C$ is smaller than the generated processing load, the expected queuing delay will eventually tend to infinity, in which case it is less interesting to examine the performance of the fog system. So we restrict our discussion to the case that generated loaded can still be processed in time, that is, $N\mu_C > N\beta\mu_L\mu_\phi$.

Theorem 4.2 *For offloading procedures that can be described by the gravity function in Eq. (4.12), the expected queuing delay $\mathbb{E}(T_{Q|n}^\epsilon) = O(\frac{1}{N})$, when $\beta\mu_L\mu_\phi < \mu_C$. The expected lifetime $\mathbb{E}(\theta_\epsilon)$ decreases as $O(\frac{1}{N})$ first, and stabilizes as $O(1)$ over N.*

Proof Assume the distributions of different parameters of the task, e.g., data volume L_δ and processing intensity ϕ_δ, are independent of each other. Then with a similar divide-and-conquer method as depicted in Fig. 4.6, we can upper bound the expected amount of data $\mathbb{E}(Y_n)$ offloaded to (arrived at) node n during time interval $[t_\epsilon, t_\epsilon + T_{S_\epsilon\to n}^\epsilon]$ by considering all the tasks generated within fog node n's communication range $g^{-1}(0)$, that is,

$$\mathbb{E}(Y_n) \leq \beta N T^{\epsilon}_{k\to n} \int_0^{g^{-1}(0)} \frac{f_V(v)}{|N_+(t_\epsilon)|} \int_0^{T^{\epsilon}_{k\to n}} \mu_\phi s H_L \left[g(v)(s - T^{\epsilon}_{k\to n})\right] \mathrm{d}s\mathrm{d}v, \tag{4.25}$$

where μ_ϕ is the expected processing intensity of tasks in Ω, and function $H_L(l) = \int_0^l x f_L(x)\mathrm{d}x$ is the partial expectation of data volume L, which is less than or equal to threshold l. Note that the denominator $|N_+(t_\epsilon)|$ follows from the upper bound of offloading probability in Eq. (4.23). Then, the expected queuing time can be upper bounded by

$$\mathbb{E}(T^{\epsilon}_{Q|n}) \leq \frac{\mathbb{E}(Y_n)}{c_n} + \left[\frac{Q_n(t_\epsilon)}{c_n} - T^{\epsilon}_{k\to n}\right]^+, \tag{4.26}$$

where $Q_n(t_\epsilon)$ is the remaining load of fog node n at time t_ϵ.

From a realistic point of view, the peak data rates in existing cellular and WiFi systems measure at 10–100 Mbps, while the processing capacity of most mobile devices measure at GHz level (e.g., 1.4 GHz Raspberry Pi 3, and 2.6~3.0 GHz smart phone processors), so the second term of Eq. (4.26) is less significant compared to the first term. Considering that $|N_+(t_\epsilon)| = \Theta(N)$, as shown in the proof of Theorem 4.1, we have $\mathbb{E}(T^{\epsilon}_{Q|n}) = O(\frac{1}{N})$ from Eqs. (4.25) and (4.26), especially when β is sufficiently large ($\beta > 0.01$). On the other hand, when β is small, i.e., the system is lightly loaded, the probability that tasks arrive at the end of transmission will be small, so $\mathbb{E}(T^{\epsilon}_{Q|n}) \simeq \max\left\{0, \frac{Y_n}{c_n} - T^{\epsilon}_{k\to n}\right\}$ will also be small positive values.

Then the expected task lifetime $\mathbb{E}(\theta_\epsilon)$ follows from the exact form of the queuing delay, as the other three delay terms in Eq. (4.24) do not change the scaling behavior over the network size N, nor are they affected by the task generation rate β. When the network size N is sufficiently large such that $\mathbb{E}(T^{\epsilon}_{Q|n})$ is small enough to be neglected, $\mathbb{E}(\theta_\epsilon)$ will remain constant over the network size N. □

Remark 4.5 It should be noted that the scaling behavior of both the queuing time and the lifetime has the same coefficient β, which is introduced by Eq. (4.25). Consequently, the $O(\frac{1}{N})$ decreasing is more evident when the system is heavily loaded, i.e., β is relatively large. Contrarily, when $\beta < 0.01$, the probability that transmissions of other offloaded tasks (to node n) complete in time interval $[t_\epsilon, t_\epsilon + T^{\epsilon}_{S_\epsilon\to n}]$ will be small, such that $\mathbb{E}(T^{\epsilon}_{Q|n}) \simeq \left[\frac{Y_n}{c_n} - T^{\epsilon}_{k\to n}\right]^+$, and the task lifetime will appear to be almost constant over the system size N.

4.4.2 Device Effort and Network Effort

With Eqs. (4.26) and (4.24), we are now able to obtain the device and network efforts for task ϵ, given that it is offloaded to node n. Consequently, an upper bound of the device effort DE_ϵ, can be obtained by $\mathbb{E}(\mathrm{DE}_\epsilon) = \sum_{i \in N_+(t_\epsilon) \cup \{k\}} \mathrm{DE}_\epsilon'' P(D_\epsilon = i)$, where the device effort DE_ϵ can be replaced by lifetime θ_ϵ, or network effort NE_ϵ.

Despite the complex form of the expected queuing delay, it outlines a generic way to foresee task ϵ's resource consumption during the offloading process, irrespective to different initial conditions and network status. Particularly in some special cases, e.g., the distance V_n follows a special distribution, we are able to obtain concrete, and more accurate results, so in the remaining of this section, we consider the special case when fog nodes follow a random walk model.

Without loss of generality, we assume the data volume L and processing intensity ϕ of tasks satisfy uniform distributions, with mean μ_L and μ_ϕ respectively. The selectiveness function takes form of $\rho(x) = x$. Consider nodes in E are initially randomly located in a square region $[-\frac{a}{2}, \frac{a}{2}]^2 \subset \mathbb{R}^2$, and they all move under a speed-constrained random walk model with a maximum speed $\mathrm{speed}_{max} << a$. When a node hits the boarder of the square region, it shows up on the opposite boarder of the region, such that there is no boarder effect [7]. In this case, the distance V between any two randomly chosen entities satisfies the following lemma.

Lemma 4.2 (Special Case of [23]) *Consider a square region* $[-\frac{a}{2}, \frac{a}{2}]^2 \subset \mathbb{R}^2$. *For two randomly chosen points* $X_1, X_2 \in [-\frac{a}{2}, \frac{a}{2}]^2$ *in this region, the distance* $V = d(X_1, X_2)$ *is a random variable with PDF as follows:*

$$f_V(v) = \begin{cases} \frac{\pi}{a^2} + \frac{v^2}{a^4}, & 0 \le v \le a, \\ -\frac{2+\pi}{a^2} + \frac{2}{a^2}\arcsin(\frac{a}{v}) + \frac{4(s^2-a^2)^{1/2}}{a^3}, & a \le v \le \sqrt{2}a. \end{cases} \tag{4.27}$$

Proof of Lemma 4.2 is very similar to the general case in [23], and is hence omitted here. With this lemma, we discuss data offloading processes under the distance-oriented and delay-oriented offloading rules. For these two offloading criteria that unifies under the generic gravity rule, we have the following theorem.

Theorem 4.3 *For distance-oriented and delay-oriented offloading, with selectiveness function* $\rho(x) = x$, *the expected network effort spent on offloading task* $\epsilon \in \Omega$ *does not change over the system size* N, *i.e.,* $\mathbb{E}_{\epsilon\in\Omega}(NE_\epsilon) = O(1)$. *The expected device effort* $\mathbb{E}_{\epsilon\in\Omega}(DE_\epsilon)$ *first decreases as* $O(\frac{1}{N})$ *to a constant value.*

Leveraging the gravity model, Theorem 4.3 outlines the scaling behavior of resource consumption during an offloading process. It indicates that, as the size N of the fog system increases, the total device effort of the system $\mathbb{E}(\sum_{\epsilon\in\Omega} \mathrm{DE}_\epsilon)$ will increase slowly (sub-linearly) at first, but rather quickly (linearly) when the size of the system exceeds a certain value. The network effort $\mathbb{E}(\sum_{\epsilon\in\Omega} \mathrm{NE}_\epsilon)$, on the other hand, increases linearly with the size of the system. In other words, there exists an optimal size of an offloading system, beyond which, the application-level delay $\mathbb{E}_\Omega(\theta_\epsilon)$ will no longer benefit from the growth of a fog system. Instead, the linearly increasing communication and coordination demand will lead to a drain of networking resource in the edge network. Next, we prove Theorem 4.3 for the distance-oriented and delay-oriented offloading scenarios separately.

4.4.2.1 Distance-Oriented Offloading

First, we discuss the simplest distance-based offloading rule, in which every task is for sure to be offloaded to a neighboring fog node, by its gravity function Eq. (4.7). For task ϵ under the this rule, the only criterion is the transmission range, i.e., $v_{max} = \frac{R_{max}}{\psi}$, because $F(\epsilon, n) > 0$ for any fog node $n \in E$.

Note that, by Lemma 4.2, the probability that two nodes happen to be located at the exact same location $F_V(v = 0)$ is strictly zero, and that it is very unlikely (with probability less than $\frac{\pi}{a^2} + \frac{1}{3a^4}$) that their distance is less than 1 unit length, especially when a is large. Therefore, the gravity force $F(\epsilon, i)$ is well-defined ($< \infty$) for almost all offloading candidates. With Lemma 4.2, we derive the lifetime and efforts of an individual task ϵ in two steps. First, we give the offloading probability, device and network effort to a fixed candidate fog node n.

Proposition 4.1 *Consider a distance-based offloading scheme with $\rho(x) = x$. For task ϵ offloaded to fog node n, the efforts spent for task ϵ are given by*

$$\mathbb{E}(\mathrm{NE}_\epsilon^n) = \frac{(1+\alpha)L_\epsilon V_n}{R_{max} - \psi V_n}, \tag{4.28}$$

$$\mathbb{E}(\mathrm{DE}_\epsilon^n) = L_\epsilon \left[\frac{(1+\alpha)L_\epsilon}{R_{max} - \psi V_n} + \frac{2\phi_\epsilon L_\epsilon}{c_n} + 2\mathbb{E}(T_{Q|n}) \right], \tag{4.29}$$

where the expected queuing time at node n is $\mathbb{E}(T_{Q|n}) \leq \frac{\beta \mu_\phi R_{max} L_\epsilon^3}{12 N c_n (R_{max} - \psi V_n)^3}$, and μ_ϕ is the mean processing intensity of all tasks.

Proof Under the distance-based gravity field, the gravity rule can be stated as $F(\epsilon, n) = \frac{1}{V_n}$ for $V_n < v_{max}$, and $F(\epsilon, n) =$ otherwise, where $v_{max} = \frac{R_{max}}{\psi}$.

Plugging in the PDF $f_V(v)$ from Lemma 4.2, we obtain the expected number of offloading candidates with a similar method in Theorem 4.1, as $\mathbb{E}(|N_+(t_\epsilon)|) = N F_V(v_{max})$, and the offloading probability can be derived as

$$\mathbb{P}(D_\epsilon = n) \simeq \left(V_n N \int_1^{v_{max}} \frac{f_V(v)}{v} \mathrm{d}v \right)^{-1} = \frac{2a^4}{N V_n \left[(v_{max})^2 + \pi a^2 \ln v_{max} \right]}, \tag{4.30}$$

where the integral in the denominator of Eq. (4.30) starts from 1 to avoid a diverging sum. We set this because the probability that the distance $V < 1$ is very small.

Then from Eq. (4.25), the expected amount of data $\mathbb{E}(Y_n)|_{T_{k\to n}^\epsilon}$ offloaded to entity n during the transmission time of task ϵ can be derived as

$$N \int_1^{v_{max}} f_V(v) \mathbb{P}(D_\epsilon = n) \beta s \mu_\phi \int_0^{T_{k\to n}^\epsilon} s H_L \left[(T_{k\to n}^\epsilon - s)(R_{max} - \psi v) \right] \mathrm{d}s \mathrm{d}v, \tag{4.31}$$

where $T^{\epsilon}_{k\to n} = \frac{L_\epsilon}{R_{max}-\psi V_n}$ is the transmission time of ϵ from the task node k to the candidate fog node n. Note that the partial expectation of data volume

$$H_L\left[(T^{\epsilon}_{k\to n} - s)(R_{max} - \psi v)\right] = \frac{1}{2}\left(L_{min} + (T^{\epsilon}_{k\to n} - s)(R_{max} - \psi v)\right), \tag{4.32}$$

because data volume L is uniformly distributed, and the lower bound L_{min} of L is small, so we can approximate this partial expectation with $\frac{1}{2}(T^{\epsilon}_{k\to n}-s)(R_{max}-\psi v)$. So we obtain $\mathbb{E}(Y_n)|_{T^{\epsilon}_{k\to n}} \simeq \frac{\beta\mu_\phi R_{max}}{12N}\left(T^{\epsilon}_{k\to n}\right)^3$.

Therefore, the expected queuing time $\mathbb{E}(T_{Q|n}) \leq \frac{\beta\mu_\phi R_{max} L_\epsilon}{12Nc_n}\left(T^{\epsilon}_{k\to n}\right)^3$. Plugging these quantities in Eq. (4.24) yields the network and device efforts. □

Proposition 4.1 gives the exact expected effort for task ϵ, if it is offloaded to fog node n, which happens with probability in Eq. (4.30). Then, we have the following corollary with respect to the expected effort spent on ϵ, in terms of network size N.

Corollary 4.1 *In distance-oriented offloading, the expected network effort* $\mathbb{E}(NE_\epsilon)$ *to offload any task* $\epsilon \in \Omega$ *is* $O(1)$ *with respect to network size* N. *The device effort* $\mathbb{E}(DE_\epsilon)$ *decreases as* $O(\frac{1}{N})$ *and converges to a constant value.*

Proof The probability that task ϵ is offloaded to fog node n is shown in Eq. (4.30), which is $O(\frac{1}{N})$. For each node n, the expected network effort for task ϵ, given that it is offloaded to n, is shown in Eq. (4.28), which is $O(1)$ with respect to the network size N. Applying the divide-and-conquer method shown in Fig. 4.6 gives us the $O(1)$ scaling law, $\mathbb{E}(|N_k(t_\epsilon)|) \sim \Theta(N)$, as shown by Eqs. (4.20) and (4.21). Then, the expected network effort for task ϵ is

$$\mathbb{E}(\text{NE}_\epsilon) = \frac{2(1+\alpha)L_\epsilon\left[(a^2\pi + \frac{R_{max}}{\psi^2})\ln R_{max} - \frac{3R_{max}^2}{2\psi^2}\right]}{\frac{R_{max}^2}{\psi} + \pi a^2\psi \ln\frac{R_{max}}{\psi}}, \tag{4.33}$$

which is $O(1)$ with respect to N.

The device effort in Eq. (4.29), on the other hand, is composed of two parts, i.e., $\frac{(1+\alpha)L_\epsilon^2}{R_{max}-\psi V_n} + \frac{2\phi_\epsilon L_\epsilon^2}{c_n}$, which is $O(1)$, and $2L_\epsilon(\mathbb{E}(T_{Q|n}))$, which is $O(\frac{\beta}{N})$. Therefore, when N is small such that $\mathbb{E}(Y_N)$ and $\mathbb{E}(T_{Q|n})$ are not negligible, the latter part dominants, and $\mathbb{E}(\text{DE}_\epsilon)$ decreases as $O(\frac{1}{N})$, especially for large β values. When β is small, or N is sufficiently large, the latter parts is negligible, so the mean device effort will not change with the network size N any more. □

Corollary 4.1 states that, the network effort and the device effort of a single task do not scale with the network size $N = |E|$. Further, if we consider all the tasks generated during the offloading process of ϵ as the total offloading traffic of the fog system, the total expected efforts, both DE and NE, scales as $O(N)$, due to the $O(N)$ number of new tasks generated per unit time. This linear growth of traffic leads to a linear growth in resource demand under the distance-oriented offloading.

4.4.2.2 Delay-Oriented Offloading

Next, we discuss the more complicated but most widely-adopted offloading criterion, i.e., delay-oriented offloading depicted by Eq. (4.9).

We adopt the same assumption as the distance-oriented offloading: there are a total number of $N = |E|$ fog nodes in the region $[-\frac{a}{2}, \frac{a}{2}]^2$, in which a is large enough, such that $a >> \max_{e \in E}\{r_e\}$. Suppose a task node k is located at the center of this region $(0, 0)$, and generates task ϵ to offload at time t_ϵ. Denote the distance from fog node n to this task node k as $V_n^\epsilon := d(X_k(t_\epsilon), X_n(t_\epsilon))$, which obeys the distribution in Lemma 4.2.

In delay-oriented offloading, whose gravity function follows the generic form, Theorem 4.1 and 4.2 still hold, so we have the following corollary, similarly as Proposition 4.1 and Corollary 4.1.

Corollary 4.2 *In delay-oriented offloading, the expected network effort to offload an individual task $\mathbb{E}(NE_\epsilon)$ is constant with respect to the size of the fog network N, and the expected device effort $\mathbb{E}(DE_\epsilon) = O(\frac{1}{N})$, which decreases first, and eventually converges to a constant value as N increases.*

Proof In delay-oriented offloading, the criterion for node n to impose a positive gravity on task ϵ $F(\epsilon, n) > 0$ in Eq. (4.15) becomes

$$F(\epsilon, n) > 0 \Leftarrow \begin{cases} C_n & > \frac{\phi_\epsilon L_\epsilon}{\tau_\epsilon}, \\ V_n^\epsilon & < \frac{1}{\psi}\left[R_{max} - \frac{(1+\alpha_\epsilon)L_\epsilon}{\tau_\epsilon - \frac{\phi_\epsilon L_\epsilon}{C_n}} \right]. \end{cases} \tag{4.34}$$

From criterion Eq. (4.34), especially the bound on distance V_n^ϵ, it can be seen that the possible offloading distance increases monotonically with the capacity of the fog node, under delay-oriented rules. Because the capacity of any fog node is uniformly distributed in $[c_{min}, c_{max}]$, the maximum distance for the task node k to choose its offloading targets can be upper bounded by plugging $C_n = c_{max}$ into Eq. (4.34). In addition, the probability that a randomly chosen fog node n qualifies as a candidate, $\mathbb{P}(F(\epsilon, n) > 0)$, can be calculated as

$$\mathbb{P}(F(\epsilon, n) > 0) = \int_{\frac{\phi_\epsilon L_\epsilon}{\tau_\epsilon}}^{c_{max}} \left. \frac{\pi a^2 v + \frac{v^3}{3}}{a^4(c_{max} - c_{min})} \right|_0^{\frac{1}{\psi}\left[R_{max} - \frac{(1+\alpha_\epsilon)L_\epsilon}{\tau_\epsilon - \frac{\phi_\epsilon L_\epsilon}{c}}\right]} dc. \tag{4.35}$$

Given that $N_k(t_\epsilon) \neq \phi$, consider a particular fog node $n \in E$ with capacity $C_n = c_n$, which is located at a distance of $V_n^\epsilon = v_n$ from the task node k. Let

$$v_{\max} = \frac{1}{\psi}\left[R_{max} - \frac{(1+\alpha_\epsilon)L_\epsilon}{\tau_\epsilon - \frac{\phi_\epsilon L_\epsilon}{c_{max}}} \right] \tag{4.36}$$

denote the maximum possible distance in Eq. (4.15) obtained by pushing C_n to its maximum value c_{max}. For an arbitrary fog node $q \in E$, when $V_q^{\epsilon} > v_{\max}$, i.e., node q is located outside the beige-shaded outer circle in Fig. 4.6, the gravity force it imposes on task ϵ will be strictly zero, such that it will never be considered as a proper candidate, i.e., $P(q \in D_\epsilon) = 0$. Therefore, it suffices to consider fog nodes with distance less than v_{max} as offloading candidates.

Similarly as in the proof of Theorem 4.1, we use the divide-and-conquer method to find the expected sum of gravity forces on task ϵ, that is, the denominator in Eq. (4.6), as

$$\mathbb{E}\left(\sum_{i\in N_k(t_\epsilon)} F(\epsilon,i)\right) = \int_{v=0}^{v_{max}} \mathbb{E}\left(\sum_{i\in N_k^v(t_\epsilon)} F(\epsilon,i)\right) \mathrm{d}v$$

$$= \int_0^{v_{max}} \frac{2\pi N v}{a^2}\left[\tau_\epsilon + \frac{(1+\alpha_\epsilon)L_\epsilon}{\psi v - R_{max}} - \frac{2\phi_\epsilon L_\epsilon}{c_{min}+c_{max}}\right]\mathrm{d}v$$

$$= \frac{2\pi N(1+\alpha_\epsilon)L_\epsilon}{\psi a^2}\ln\frac{R_{max}-\psi v_{max}}{R_{max}} + \frac{\pi N}{a^2}\left[\tau_\epsilon v_{max}^2 - \frac{4\phi_\epsilon L_\epsilon v_{max}}{c_{min}+c_{max}}\right]. \tag{4.37}$$

From Eq. (4.37), we know that the probability that task ϵ is offloaded to fog node n is $O(\frac{1}{N})$, as the numerator in definition Eq. (4.6) is a constant over N. Then for each node n, the expected effort for task ϵ, given that the task is offloaded to n, are also $O(1)$ with respect to N. Applying the divide-and-conquer method shown in Fig. 4.6, as that in the proof of Theorems 4.1 and 4.2, we obtain the same scaling behavior as that stated in Corollary 4.1, because $\mathbb{E}(|N_k(t_\epsilon)|) \sim \Theta(N)$ and $\mathbb{E}(\theta_\epsilon) = O(\frac{1}{N})$ hold for the generic gravity function, including delay-oriented offloading.

□

4.5 Numerical Results and Discussions

The significance of the proposed gravity model is twofolded. First, as a versatile model, it can be flexibly configured to mimic the behavior of different offloading schemes, such that their performances can be evaluated through analysis or simulation to answer the 'remedy or resource drain' question for massive offloading in large-scale systems. On the other hand, the gravity model itself can also be executed as a distributed offloading scheme on task nodes, whose performance can be compared with others to select the most appropriate scheme for IoT applications.

To validate these two aspects for the proposed gravity model, we simulate task offloading in fog systems of different sizes, with configurations listed in Table 4.4.

Specifically, we consider a $1000\times1000\,\text{m}^2$ region, where task/fog nodes move with a speed-constrained random walk of maximum speed 10 m/s. The maximum data rate R_{max} is set to be 10 Gbps, according to the RAT configuration listed

Table 4.4 Simulation configuration

Parameter	Description	Value/Range
$N = \|E\|$	Network size	100, 500, 1000, 5000, 10,000
a	Edge length of the region	1000 m
N/A	Simulation run times	100
N/A	Number of tasks	1000
β	Task generation rate	0.01, 0.05, 0.5
$speed_{max}$	Max moving speed	10 m/s
ψ	Data rate attenuation	20 Mbps/m
R_{max}	Max data rate	10 Gbps
L	Data volume of tasks	1 Mb, 10 Mb, 100 Mb, 1 Gb
ϕ	Processing intensity of tasks	100, 200, 300, 400 cycles/bit
α	Volume change coefficient of tasks	0.1, 0.5, 1
C	CPU frequency of fog nodes	1 GHz (w.p. 0.4), 4 GHz (w.p. 0.6)
P_e	Transmission power	[0.01, 0.25] W
ξ	Processing power	$[1.3, 3.5]\times10^{-7}$ J/cycle

in Table 4.2. We consider tasks with different data volumes, ranging from 1 Mb to 1 Gb, accounting for data collection, translation, video trans-coding, VR/AR applications. This uniform distribution with uneven dispersion is motivated by existing IoT application traffics[32], future VR/AR application traffic [2, 25], also following conventions of related works[10, 18, 27, 28]. The number of task/fog nodes ranges from 100 to 10,000, while the task generation rate β varies from 0.01 to 0.5, to examine the performances as the system scales. Transmission power of task/fog nodes ranges from 0.01 to 0.25 W, taking different RAT's in Table 4.2 into consideration. The processing power of fog nodes ranges from 1.3 to 3.5×10^{-7} J/cycle, referencing [28]. For each configuration, the simulation is run for 100 times, each of which has randomly changing initial configurations (light colored markers), to obtain the ensemble mean (bright colored markers and lines).

4.5.1 Describing an Offloading Scheme with the Gravity Model

We first examine whether the gravity model can describe offloading schemes with different offloading criteria. Without loss of generality, we inspect the delay-oriented gravity model, as delay is the primary concern in fog-based IoT.

From the perspective of fog nodes, the CPU occupancy in Fig. 4.7 quantifies the percentage of time slots in which the CPU of a fog node is busy processing tasks. The $O(\frac{1}{N})$ decreasing occupancy over the number of nodes N echoes with the analysis of offloading probability in Theorem 4.1 and the offloaded processing load $\mathbb{E}(Y_n)$ in Theorem 4.2. Meanwhile, relaxing the selectiveness function from $\rho(x) = x^{10}$ (blue dashed line with triangle markers) to $\rho(x) = x$ (red dashed line

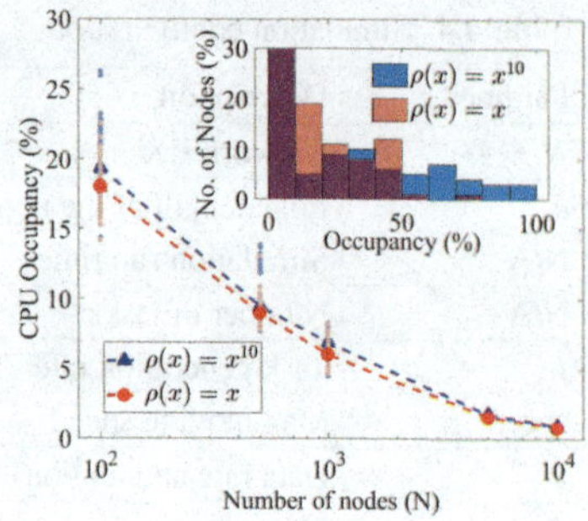

Fig. 4.7 CPU occupancy decreases with the system size N. Moreover, relaxing the selectiveness (from $\rho(x) = x^{10}$ to $\rho(x) = x$) achieves load-balancing among fog nodes

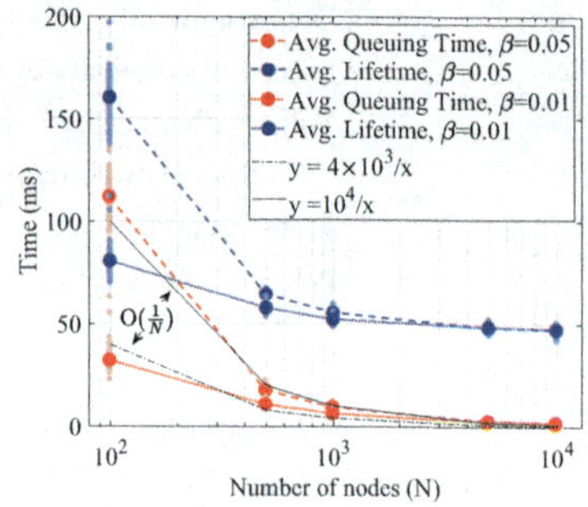

Fig. 4.8 Both the queuing delay $\mathbb{E}(T^{\epsilon}_{Q|n})$ and task lifetime $\mathbb{E}(\theta_{\epsilon})$ of an individual task are proportionate to the task generating rate β, and exhibit $O(\frac{1}{N})$ decreasing with respect to the network size N

with round markers) successfully balance the processing load among fog nodes, as indicated by the lower average CPU occupancy as well as the number of high-occupancy nodes in the inner plot. When $\rho(x) = x$, task nodes are less reluctant to switch offloading targets, as a result of which, the processing load is divided more evenly among fog nodes (red bars), instead of overloading a small number of powerful fog nodes (blue bars that concentrating at the high occupancy end). From the perspective of the offloading service, the task queuing delay (red lines and markers) and task lifetime (blue lines and markers) in Fig. 4.8 are both $O(\frac{1}{N})$, as stated by Theorem 4.2. These indicate the designed gravity model can describe a family of offloading schemes.

4.5.2 *Comparison with Existing Offloading Schemes*

Next, to examine the detailed behavior of the gravity model, which is a fully distributed scheme that does require any coordination, we compare it with the cooperative Myopic Best Response (MBR) offloading [18] and Matching Theory-based Efficient Task Offloading (METO) [28], to test offloading scenarios with single criterion (delay-oriented) and multiple criteria (delay-energy-hybrid), respectively.

4.5.2.1 Single Criterion

The MBR scheme [18] is designed to minimize the task lifetime, so we compare it with the gravity model with delay-oriented gravity function in Eq. (4.9). Numerical results of this comparison are presented in Figs. 4.9, 4.10, and 4.11. MBR is expected to achieve the global optimal (at the current time slot) with necessary information of all fog nodes, such as the queue length on each candidate fog node, so an extra communication time (2 ms and 10 ms) is added in the simulation as overhead. We highlight the following three observations from the comparison.

1. *Overlapping Offloading Targets*: The target in MBR is usually the same candidate fog node that imposes the largest gravity force on the task, with a hit ratio over 70%, as shown in Fig. 4.9. When it is not the case, the MBR target still resides in the top 3 candidates obtained by the gravity model.

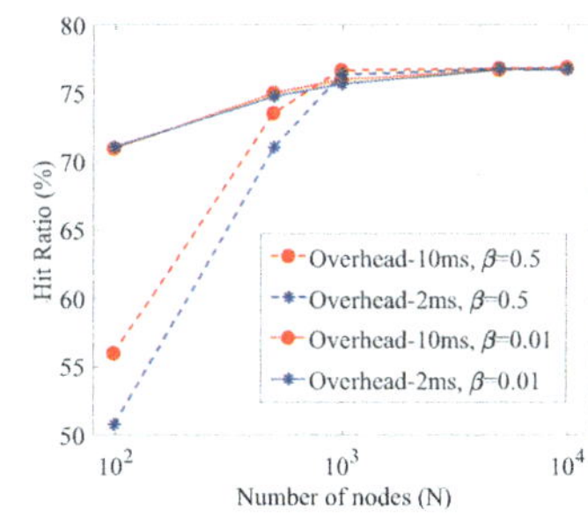

Fig. 4.9 In over 50% of the offloading procedures, the MBR target fog node also imposes the largest gravity

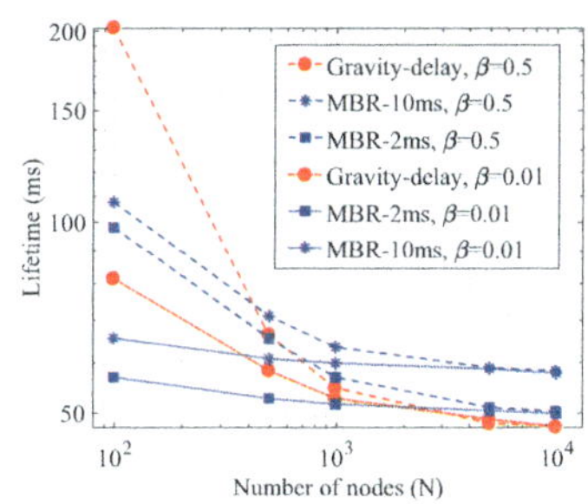

Fig. 4.10 The delay-oriented gravity model achieves lower application delay than MBR

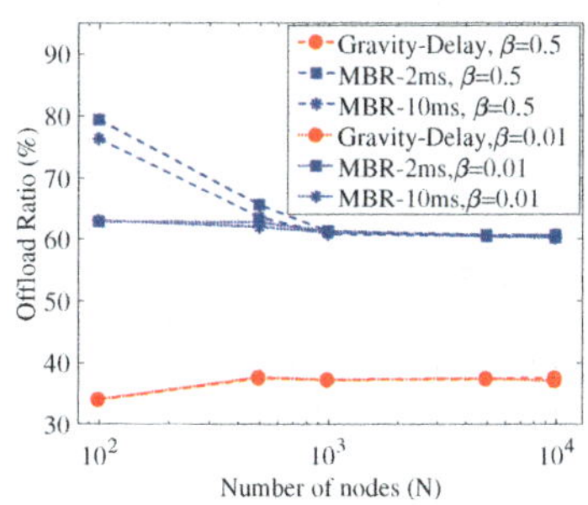

Fig. 4.11 The gravity model has a higher preference of local processing, which may cause the difference in ranking the offloading candidates

2. *Similar but Better Delay Performance*: In both schemes, the decreasing trends of the task lifetime in Fig. 4.10, and hence the device effort, are all $O(\frac{1}{N})$ as the system scales, indicating that the gravity model accurately captures the performance change in the fog-based IoT system. Despite the same $O(\frac{1}{N})$ decreasing trend, the gravity model, which is also a distributed offloading scheme, achieves a lower task lifetime than the MBR scheme in most scenarios. The reason behind this delay reduction is the lower queuing delay due to the automatic load-balancing, which is especially true when the system is large. In these cases, the gravity model allows a strictly positive probability to offloading a task to the 'sub-optimal' candidate to avoid overloading some fog nodes.
3. *Diverging Offloading Ratios and Network Efforts*: The main difference between the two offloading schemes lies in the offloading ratio, as shown in Fig. 4.11, leading to a higher average network effort. The gravity model tends to execute tasks locally, due to its strict criterion for offloading, resulting in a lower than 50% offloading ratio. However, this ratio can be tuned by adjusting the threshold (zero in this case) of triggering an offloading, as will be discussed next.

4.5.2.2 Multiple Criteria

The METO scheme [28] allows a task node to employ multiple offloading criteria, but requires global information, including processing time and energy, to rank the candidate and perform matching. We compared it with the gravity model in a delay-energy-hybrid offloading scenario, where half of the tasks are subject to delay-oriented offloading and the other half to energy-oriented offloading. To better compare the two, we combine MBR and the TOPSIS-ranking algorithm in METO to enhance its performance, such that the queuing time is also taken into consideration for the delay-oriented offloading criterion. Numerical results are presented in Figs. 4.12, 4.13, and 4.14, by which we observe that the offloading ratio and energy consumption of task nodes are well captured by the gravity model. We also observe that the MBR-enhanced METO achieves a task lifetime as low as the delay-oriented offloading with gravity model. This gap in total energy consumption and task lifetime can be attributed to the full information of candidate fog nodes (which incurs communication overheads of at least $O(N)$ rounds of control message exchanges that are not accounted for), and the TOPSIS-based ranking calculation, whose complexity is $O(N^2)$. Nonetheless, the scaling behavior of energy consumption and task lifetime over the system size N are still captured by the gravity model.

Results in Figs. 4.7, 4.8, 4.9, 4.10, 4.11, 4.12, 4.13, and 4.14 indicate that the proposed gravity model can effectively describe an offloading process, while the designed forms of gravity and selectiveness allow flexible adjustments in offloading preferences. Moreover, it also enables us to theoretically analyze the performance of offloading service in the fog-based IoT, as these analytical results apply to systems of any scale (N).

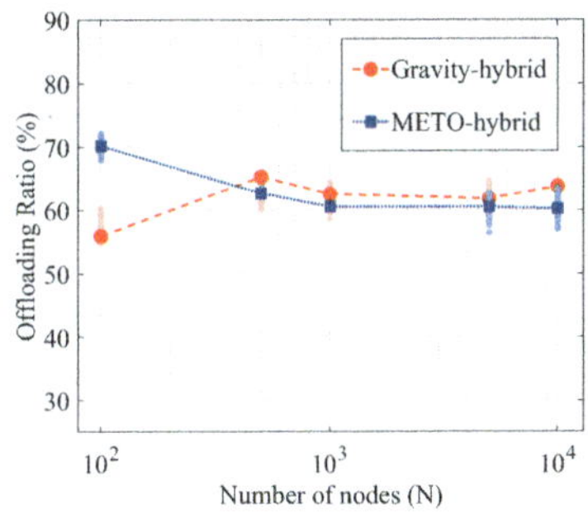

Fig. 4.12 The offloading ratio of METO is well captured by the gravity model, especially when the system (N) is large

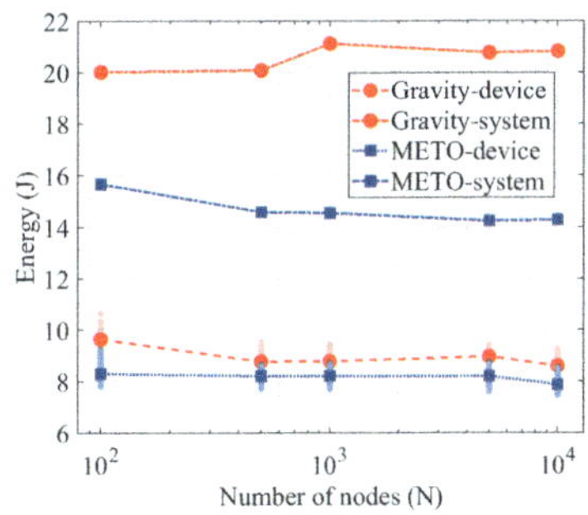

Fig. 4.13 The gravity model achieves similar local energy consumption, i.e., energy consumed by the task node, as METO, but consumes more energy in total

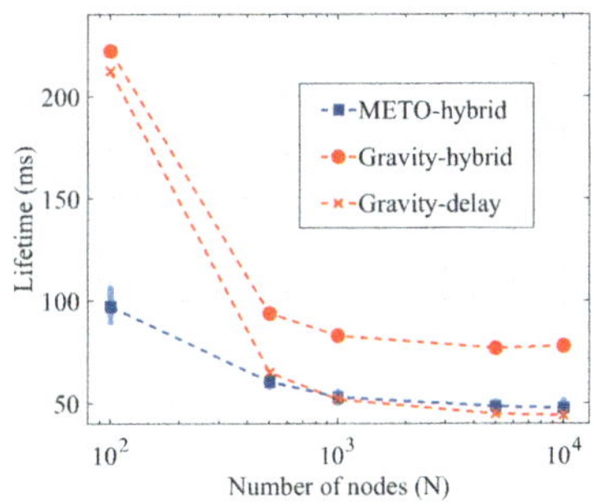

Fig. 4.14 The scaling of the lifetime is captured, but METO achieves lower lifetime due to the MBR+TOPSIS enhancement

4.5.3 *Gravity Model as an Offloading Scheme*

Considering its advantages of being fully distributed, and easy for large-scale analysis, the gravity model itself can also be implemented as an offloading scheme. To further examine its performances in different scenarios, more simulations are conducted under different task generation rates (β) and selectiveness functions ($\rho(x)$), as shown in Figs. 4.15, 4.16, and 4.17. We highlight the following observations from these results.

1. *Theorems 4.2 and 4.3 hold for all cases*. The scaling laws over the network size N derived in Theorem 4.2 and 4.3, i.e., $O(\frac{1}{N})$ task lifetime, $O(\frac{1}{N})$ device effort, and $O(1)$ network effort, are captured by the red lines in Fig. 4.15.

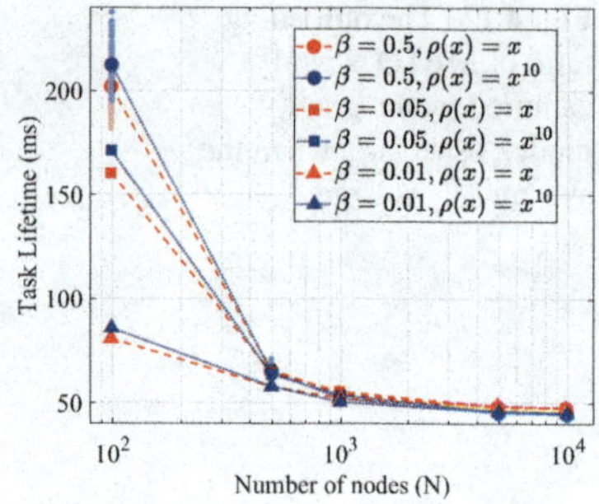

Fig. 4.15 Under the gravity offloading model, the task lifetime scales as $O(\frac{1}{N})$

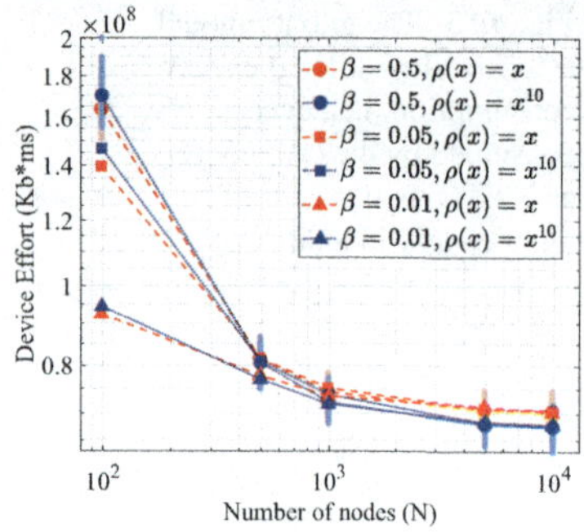

Fig. 4.16 Under the gravity offloading model, the device effort scales as $O(\frac{1}{N})$

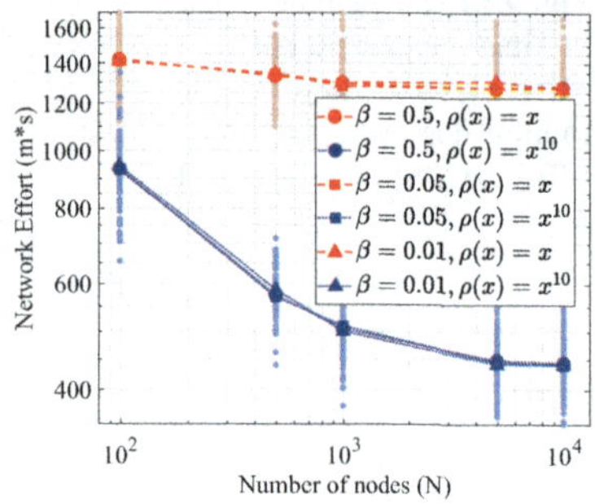

Fig. 4.17 Under the gravity offloading model, .the network effort scales as $O(1)$

2. *Rate β does not change the scaling behavior*. Theorems 4.2 and 4.3 hold for all task generation rates (β). The higher the processing load (larger β), the longer the application-level delay, and the larger device effort. Due to its stochastic nature, however, the amount of communication resource spent on a single task does not change over the workload, as indicated by the overlapping red/blue lines in Fig. 4.17). This behavior differs from the MBR scheme, but is especially useful in network planning due to its insensitivity to processing load.
3. *Selectiveness $\rho(x)$ greatly affects the network effort*. Theorem 4.2 holds for all $\rho(x)$ in polynomial forms, which is well captured by both the red and blue lines in Fig. 4.15. Consequently, device effort also follows the $O(\frac{1}{N})$ decreasing, as shown in Fig. 4.16. The network effort, nonetheless, is greatly affected by $\rho(x)$, changed from $O(1)$ to $O(\frac{1}{N})$ as indicated by the blue lines in Fig. 4.17,

alleviating the possibility of resource drain when the system scales. Moreover, we see that tightening selectiveness in the gravity model to $\rho(x) = x^{10}$ can capture the decreasing trend of network effort of the MBR scheme and the METO scheme, implying the flexibility and descriptiveness of the gravity model.

From the perspective of performance, these simulation results indicate that, task offloading in the fog system is a remedy to delay-sensitive applications for IoT, as task lifetime decreases quickly with the size of the network. Nonetheless, it is also a possible resource drain to the edge network, in the sense that the linearly growing consumption of total communication resource, i.e., the sum of network effort during a certain time period, may not be sustained in this course, considering the limited network capacity of the edge network. While resource drain at the edge may be attributed to many possible reasons, including spectrum misuse, network overlay, and so on, uncontrolled, massive offloadings introduced by an open fog system clearly adds to the risk, especially considering the explosive growth of IoT devices.

4.6 Summary

In this chapter, we discuss the performance evaluation problem for task offloading processes in the fog, through the lifetime of individual tasks, device effort, and network effort defined via data movements. We propose a gravity-based offloading model, by which a variety of offloading criteria can be described. The storage and communication resource demand are then quantified by the device effort and network effort metrics, such that resource consumption under different schemes can be measured through a set of unified metrics. Through analysis and simulation, we find that the time and efforts to offload a single task do not scale with the network size, which indicates that, the total resource consumption for all the data offloading processes scale linearly with the network size, under the gravity-based offloading schemes. As our model may be employed to describe various offloading criteria, these results are generally applicable in understanding how multiple tasks move in a resource-constrained system, such as the fog paradigm, as well as the impact (in the forms of resource consumption) on the underlying provision networks.

References

1. Ai, Y., Peng, M., Zhang, K.: Edge computing technologies for internet of things: a primer. Digital Commun. Netw. **4**(2), 77–86 (2018). https://doi.org/10.1016/j.dcan.2017.07.001. http://www.sciencedirect.com/science/article/pii/S2352864817301335
2. Al Mtawa, Y., Haque, A., Bitar, B.: The mammoth internet: are we ready? IEEE Access **7**, 132894–132908 (2019). https://doi.org/10.1109/ACCESS.2019.2941110

3. Aparicio-Pardo, R., Pires, K., Blanc, A., Simon, G.: Transcoding live adaptive video streams at a massive scale in the cloud. In: Proceedings of the 6th ACM Multimedia Systems Conference, MMSys '15, pp. 49–60. ACM, New York (2015). https://doi.org/10.1145/2713168.2713177. http://doi.acm.org/10.1145/2713168.2713177
4. Baek, J., Kaddoum, G.: Heterogeneous task offloading and resource allocations via deep recurrent reinforcement learning in partial observable multifog networks. IEEE Internet Things J. **8**(2), 1041–1056 (2021). https://doi.org/10.1109/JIOT.2020.3009540
5. Baek, J., Kaddoum, G.: Online partial offloading and task scheduling in sdn-fog networks with deep recurrent reinforcement learning. IEEE Internet Things J. **9**(13), 11578–11589 (2022). https://doi.org/10.1109/JIOT.2021.3130474
6. Baek, J.Y., Kaddoum, G., Garg, S., Kaur, K., Gravel, V.: Managing fog networks using reinforcement learning based load balancing algorithm. In: 2019 IEEE Wireless Communications and Networking Conference (WCNC), pp. 1–7 (2019). https://doi.org/10.1109/WCNC.2019.8885745
7. Bai, F., Helmy, A.: A survey of mobility models. Wirel. Adhoc Netw. **206**, 147 (2004)
8. Baylis, C., Fellows, M., Cohen, L., Marks II, R.J.: Solving the spectrum crisis: Intelligent, reconfigurable microwave transmitter amplifiers for cognitive radar. IEEE Microw. Mag. **15**(5), 94–107 (2014). https://doi.org/10.1109/MMM.2014.2321253
9. Borges, R.M., Pereira, L.A.M., Ferreira, A.C., Mendes, L.L., Spadoti, D.H., Cerqueira Sodre, Jr., A.: Fifth-generation new radio fiber-wireless system for long-reach and enhanced mobile broadband scenarios. Microw. Opt. Technol. Lett. **63**(2), 662–669 (2021)
10. Chen, M., Hao, Y.: Task offloading for mobile edge computing in software defined ultra-dense network. IEEE J. Sel. Areas Commun. **36**(3), 587–597 (2018)
11. Chiti, F., Fantacci, R., Picano, B.: A matching theory framework for tasks offloading in fog computing for iot systems. IEEE Internet Things J. **5**(6), 5089–5096 (2018). https://doi.org/10.1109/JIOT.2018.2871251
12. Cisco Visual Networking Index: Forecast and Trends, 2017–2022 White Paper. Technical Report, Cisco and/or its affliates (2019)
13. Fan, Q., Bai, J., Zhang, H., Yi, Y., Liu, L.: Delay-aware resource allocation in fog-assisted iot networks through reinforcement learning. IEEE Internet Things J. **9**(7), 5189–5199 (2022). https://doi.org/10.1109/JIOT.2021.3111079
14. Fogg, I.: The State of Wifi vs Mobile Network Experience as 5G Arrives. Technical Report, OpenSignal (2018)
15. He, J., Wei, J., Chen, K., Tang, Z., Zhou, Y., Zhang, Y.: Multitier fog computing with large-scale iot data analytics for smart cities. IEEE Internet Things J. **5**(2), 677–686 (2018). https://doi.org/10.1109/JIOT.2017.2724845
16. Huo, Y., Dong, X., Xu, W.: 5g cellular user equipment: From theory to practical hardware design. IEEE Access **5**, 13992–14010 (2017). https://doi.org/10.1109/ACCESS.2017.2727550
17. IEEE Std 1934-2018: Ieee standard for adoption of openfog reference architecture for fog computing, pp. 1–176 (2018). https://doi.org/10.1109/IEEESTD.2018.8423800
18. Josilo, S., Dan, G.: Decentralized algorithm for randomized task allocation in fog computing systems. IEEE/ACM Trans. Netw. **27**(1), 85–97 (2019). https://doi.org/10.1109/tnet.2018.2880874
19. Liu, Z., Yang, X., Yang, Y., Wang, K., Mao, G.: Dats: Dispersive stable task scheduling in heterogeneous fog networks. IEEE Internet Things J. **6**(2), 3423–3436 (2019). https://doi.org/10.1109/JIOT.2018.2884720
20. Lottarini, A., Ramirez, A., Coburn, J., Kim, M.A., Ranganathan, P., Stodolsky, D., Wachsler, M.: vbench: Benchmarking video transcoding in the cloud. In: ACM SIGPLAN Notices, vol. 53, pp. 797–809. ACM (2018)
21. Mao, Y., You, C., Zhang, J., Huang, K., Letaief, K.B.: A survey on mobile edge computing: The communication perspective. IEEE Commun. Surv. Tutor. **19**(4), 2322–2358 (2017). https://doi.org/10.1109/COMST.2017.2745201

22. Perera, C., Liu, C.H., Jayawardena, S.: The emerging internet of things marketplace from an industrial perspective: A survey. IEEE Trans. Emerg. Top. Comput. **3**(4), 585–598 (2015). https://doi.org/10.1109/TETC.2015.2390034
23. Philip, J.: The probability distribution of the distance between two random points in a box. KTH mathematics, Royal Institute of Technology (2007)
24. Pu, L., Chen, X., Xu, J., Fu, X.: D2d fogging: An energy-efficient and incentive-aware task offloading framework via network-assisted d2d collaboration. IEEE J. Sel. Areas Commun. **34**(12), 3887–3901 (2016). https://doi.org/10.1109/JSAC.2016.2624118
25. Qualcomm: VR and AR pushing connectivity limits. Technical Report (2018)
26. Rahman, M.A., Hossain, M.S., Hassanain, E., Muhammad, G.: Semantic multimedia fog computing and iot environment: Sustainability perspective. IEEE Commun. Mag. **56**(5), 80–87 (2018). https://doi.org/10.1109/MCOM.2018.1700907
27. Shi, J., Du, J., Wang, J., Wang, J., Yuan, J.: Priority-aware task offloading in vehicular fog computing based on deep reinforcement learning. IEEE Trans. Veh. Technol. **69**(12), 16067–16081 (2020). https://doi.org/10.1109/TVT.2020.3041929
28. Swain, C., Sahoo, M.N., Satpathy, A., Muhammad, K., Bakshi, S., Rodrigues, J.J.P.C., de Albuquerque, V.H.C.: Meto: Matching-theory-based efficient task offloading in iot-fog interconnection networks. IEEE Internet Things J. **8**(16), 12705–12715 (2021). https://doi.org/10.1109/JIOT.2020.3025631
29. Tang, L., He, S.: Multi-user computation offloading in mobile edge computing: a behavioral perspective. IEEE Netw. **32**(1), 48–53 (2018). https://doi.org/10.1109/MNET.2018.1700119
30. Tong, L., Li, Y., Gao, W.: A hierarchical edge cloud architecture for mobile computing. In: IEEE INFOCOM 2016 - The 35th Annual IEEE International Conference on Computer Communications, pp. 1–9 (2016). https://doi.org/10.1109/INFOCOM.2016.7524340
31. Vaquero, L.M., Rodero-Merino, L.: Finding your way in the fog: Towards a comprehensive definition of fog computing. SIGCOMM Comput. Commun. Rev. **44**(5), 27–32 (2014). https://doi.org/10.1145/2677046.2677052. http://doi.acm.org/10.1145/2677046.2677052
32. Wan, Y., Xu, K., Wang, F., Xue, G.: Characterizing and mining traffic patterns of iot devices in edge networks. IEEE Trans. Netw. Sci. Eng. **8**(1), 89–101 (2021). https://doi.org/10.1109/TNSE.2020.3026961
33. Wang, T., Zhou, J., Liu, A., Bhuiyan, M.Z.A., Wang, G., Jia, W.: Fog-based computing and storage offloading for data synchronization in iot. IEEE Internet Things J. 1–1 (2019). https://doi.org/10.1109/JIOT.2018.2875915
34. Wu, H.: Performance modeling of delayed offloading in mobile wireless environments with failures. IEEE Commun. Lett. **22**(11), 2334–2337 (2018). https://doi.org/10.1109/LCOMM.2018.2866585
35. Xiao, Y., Krunz, M.: Qoe and power efficiency tradeoff for fog computing networks with fog node cooperation. In: IEEE INFOCOM 2017 - IEEE Conference on Computer Communications, pp. 1–9 (2017). https://doi.org/10.1109/INFOCOM.2017.8057196
36. Yang, Y., Chang, X., Han, Z., Li, L.: Delay-aware secure computation offloading mechanism in a fog-cloud framework. In: 2018 IEEE Intl Conference on Parallel Distributed Processing with Applications, Ubiquitous Computing Communications, Big Data Cloud Computing, Social Computing Networking, Sustainable Computing Communications (ISPA/IUCC/BDCloud/SocialCom/SustainCom), pp. 346–353 (2018). https://doi.org/10.1109/BDCloud.2018.00061
37. Yao, J., Ansari, N.: Reliability-aware fog resource provisioning for deadline-driven iot services. In: 2018 IEEE Global Communications Conference (GLOBECOM), pp. 1–6 (2018). https://doi.org/10.1109/GLOCOM.2018.8647378
38. Yao, J., Ansari, N.: Task allocation in fog-aided mobile iot by lyapunov online reinforcement learning. IEEE Trans. Green Commun. Netw. **4**(2), 556–565 (2020). https://doi.org/10.1109/TGCN.2019.2956626
39. Yi, S., Li, C., Li, Q.: A survey of fog computing: concepts, applications and issues. In: Proceedings of the 2015 Workshop on Mobile Big Data, Mobidata '15, pp. 37–42. ACM, New York (2015). https://doi.org/10.1145/2757384.2757397

40. Yousefpour, A., Fung, C., Nguyen, T., Kadiyala, K., Jalali, F., Niakanlahiji, A., Kong, J., Jue, J.P.: All one needs to know about fog computing and related edge computing paradigms: a complete survey. J. Syst. Archit. (2019). https://doi.org/https://doi.org/10.1016/j.sysarc.2019.02.009. http://www.sciencedirect.com/science/article/pii/S1383762118306349
41. Zhang, H., Xiao, Y., Bu, S., Niyato, D., Yu, F.R., Han, Z.: Computing resource allocation in three-tier iot fog networks: a joint optimization approach combining stackelberg game and matching. IEEE Internet Things J. **4**(5), 1204–1215 (2017). https://doi.org/10.1109/JIOT.2017.2688925

Chapter 5
Spectrum Dynamics: Modeling, Analysis, and Design of Spectrum Activity Surveillance in DSA-Enabled Systems

In the previous chapter, we discussed the impact on network resource by data movements in the offloading process, but it is still not clear *how to observe* such impact. This is especially true for the spectrum resource, which is already scarce in existing wireless systems. On the other hand, future wireless systems are expected to host a large number of users with a variety of RAT, including dynamic spectrum access dynamic spectrum access (DSA), the coordination among which requires better understanding of instantaneous spectrum occupancy status. In this chapter, we study spectrum activity surveillance (SAS) to address *what is the observable impact of mobile data*. Due to the growing concern on spectrum efficiency, SAS is essential to dynamic spectrum access (DSA)-enabled wireless systems. As monitoring hardware is actively being developed, we take a modeling approach toward this question, such that the monitor model can be fine-tuned to describe spectrum monitors with different characteristics.

We introduce a three-factor space, composed of *spectrum*, *time*, and *geographic region*, in which spectrum activities are only observable in a *closed* subspace, due to their locality in the three domains. We identify and quantify the two objectives of SAS, based on which we formulate the strategy design problem in two steps: 3D-tessellation for sweep (monitoring) *coverage* and graph walk for detecting *spectrum culprits*, that is, wireless devices responsible for unauthorized spectrum occupancy. To observe the spectrum dynamics as the result of mobile data, we design efficient SAS strategies with multiple monitors: a low-switching-cost strategy for systems with dedicated monitors, and performance-guaranteed random strategies for systems with crowdsourcing monitors. We find that randomized strategies by m monitors (even with limited switching capacities) can achieve a sweep coverage over a space of n assignment points in $\Theta(\frac{n}{m}\ln n)$ time, and detect an oblivious or adversarial spectrum culprit in $\Theta(\frac{n}{m})$ time. This $O(\frac{1}{m})$ scaling indicates that, both SAS objectives can be achieved with a linear 'speedup' as more spectrum monitors can be deployed, which offers performance expectations for existing systems. In addition, size n of the assignment space is obtained as a

J. Wang et al., *Encountering Mobile Data Dynamics in Heterogeneous Wireless Networks*, https://doi.org/10.1007/978-3-031-62906-8_5

function of the monitoring power of individual monitors, so the performance bounds in turn outline design requirements for spectrum monitors, in order to achieve a certain level of SAS performance.

5.1 Motivation and Related Work

Dynamic Spectrum Access (DSA) has been envisioned as one of the key technologies for high-speed heterogeneous wireless systems [14], e.g., 5G cellular networks [29], since it is expected to boost spectrum efficiency by allowing wireless devices to temporally operate beyond their designated spectrum bands, so as to mitigate the gap between the increasing frequency demand and the crowded licensed radio spectrum. It is important on both individual and system levels: it is essential to advanced cognitive radio (CR) technologies, e.g., CR non-orthogonal multiple access (CR-NOMA) [11]; and it is also preliminary to abstraction of wireless resources in a system, e.g., wireless network virtualization (WNV) [19, 21]. Despite its great potentials, the opportunistic and open nature of DSA bears an intrinsic demand for *spectrum activity surveillance (SAS)*, as both a prerequisite and a supplement to such spectrum-agile systems.

5.1.1 Motivation

A SAS process is expected to carry out continuous scan of spectrum activities on the frequencies of interest, for the purpose of usage data collection and spectrum regulation policing/enforcing. On a systematic level, surveillance logs reflect the spectrum usage in wireless communication systems, and can hence be analyzed for system management, as well as data disclosure [13] purposes; on an individual level, real-time spectrum occupancy status can serve as a crude input to reveal and predict the spectrum sensing range [23] for opportunistic spectrum access, which can reduce the sensing time as well as the access delay. To this end, Google [12] and Microsoft [22] have launched their *spectrum database* projects, which provide availability of TV white space over the entire United States, as a preliminary step toward real-time DSA applications and the construction of *radio environment map (REM)* [15]. In this sense, SAS is expected to act as a 'spectrum-meter', passively recording instantaneous occupancy data of different spectrum slices for future analysis.

On the other hand, as an immediate beneficiary of this opportunistic environment toward higher spectrum efficiency, spectrum *culprit*, which refers to overly-aggressive or malicious users, may undermine the 'right-of-way' of legitimate users, and even downgrading performance of the entire system, by occupying unauthorized frequency bands that are promised to other legitimate users. This problem is especially severe in DSA-enabled systems with distributed spectrum sharing schemes, where a simple Listen-Before-Talk (LBT) mechanism [32] is preferred due to its scalability and comparable throughput performances. In such systems, it is easy for 'smart' spectrum culprits to abuse the DSA-enabled system, owing to the application of machine learning in cognitive radios [24, 30]. Consequently, SAS is expected to act as the 'spectrum-police', pro-actively detecting spectrum misuse, guarding the rights of legitimate users, and preserving forensics for further actions.

Therefore, SAS is both a premise to leverage spectrum efficiency in compliance to policy enforcement, and a proactive approach to catch the spectrum culprits. Such a system-level function of a DSA-enabled system is completed by *spectrum monitors*, who take advantage of spectrum sensing, networking, and data processing techniques, to collect spectrum occupancy measurements, and identify spectrum culprits based on collected data. In other words, a spectrum monitor is logically composed of three building blocks: sensing hardware, measurement/detection algorithm, and communication protocol. For a large-scale commercial DSA-enabled system, e.g., a multi-operator LTE-WiFi overlay network in the 5 GHz unlicensed frequency bands described in [32], it is necessary to include, and coordinate multiple monitors to provide reliable and timely-updated SAS results.

5.1.2 Related Work

In this regard, existing literature on SAS can be broadly summarized into two categories: single-monitor technique and multiple-monitor orchestration.

The former develops prototypes [23, 25], technique, and algorithms [20, 34] for individual spectrum monitors, that can effectively differentiate spectrum misuse or abnormalities from normal activities, e.g., statistical significance testing [20], and the spectrum permit mechanism [34], for security enhancement and attack mitigation. In contrast, the latter focuses on efficient deployment and cooperation of *multiple* monitors for the purpose of better surveillance coverage [27], lower switching cost [17], or faster detection of culprits [18]. To this end, spectrum occupancy measurement and interference map construction with commercial dedicated monitors have been studied in [15], while the crowdsourcing sensing/monitoring paradigm is proposed for cost and flexibility improvement, taking advantage of collaboration [17] and distributed data decoding [6].

In prior studies of multiple monitor deployment strategies, e.g., [17, 18, 27], an implicit assumption is made for spectrum monitors to be sufficiently powerful, such that they can watch over the entire geographical region of interest and tune/move without any limit. The fact, however, is that most spectrum activities, including communications, attacks/jamming and monitoring/sniffing, are *local*, i.e., confined in both the frequency domain and the space domain during a fixed-length time interval, as noted in prototype design [25], and spectrum occupancy measurements [15]. This discrepancy is especially pronounced in wide-band wide-area monitoring, e.g., spectrum database or radio environment map (REM) construction, which naturally leads to an open question: *how to perform an spectrum activity surveillance (SAS)) process and design SAS strategies (with multiple monitors) for DSA-enabled systems?*

5.1.3 Our Approach and Contributions

Hindered by the constraints on spectrum license and high deployment expenses, studying the SAS problem via field tests is not a viable option, especially at the early stage when development of prototypes [23, 25], as well as standardization for CR and DSA, are still underway. Therefore, taking into consideration various monitor settings and SAS scenarios, we take a modeling approach to study SAS processes from perspectives of surveillance coverage and culprit detection. Seemingly trivial, the SAS problem is actually challenging in the following aspects. First, objectives of SAS, such as data collection and culprit detection, are by-and-large global and collective, lacking a consolidated measure, through which a monitor deployment strategy can be fairly evaluated. Second, if spectrum is considered as an 1-D domain, the surveillance problem over a geographical region is naturally extended to a 3-D space, in which tracking surveillance coverage and analysis are both non-trivial.

To address these challenges, we construct a *spectra-location space* that incorporates spectra, temporal and geographical domains, in which the locality of spectrum activities are captured by limited range and closed spaces. With respect to the SAS problem, our contributions can be summarized as follows:

We formally define *monitoring power*, *switching cost*, and *switching capacity* to characterize monitors' and culprits' spectrum activities, and formulate the SAS process into a tractable graph walk process with space-tessellation, such that a collective surveillance function is transformed into localized (even distributed) actions of individual monitors.

We translate the qualitative data collection and culprit detection objectives of SAS processes into two quantitative metrics in the time domain, i.e., the *coverage time* and *detection time*, such that different SAS (monitors deployment) strategies can be evaluated, and fairly compared.

We also present a deterministic SAS strategy with low switching cost for systems with dedicated spectrum monitors, and randomized strategies specialized to protect against adversarial spectrum culprits, which is suitable for crowdsourcing surveillance scenarios. Despite the switching capacity limit, randomized strategies of m monitors can achieve a full sweep coverage over a spectra-location space of n assignment points in $\Theta(\frac{n}{m} \ln n)$ time, and detect a persistent or adversarial culprit in $\Theta(\frac{n}{m})$ time.

5.2 Problem Formulation

In this section, we formally define the spectra-location space, spectrum activities and performance metrics to formulate the SAS problem.

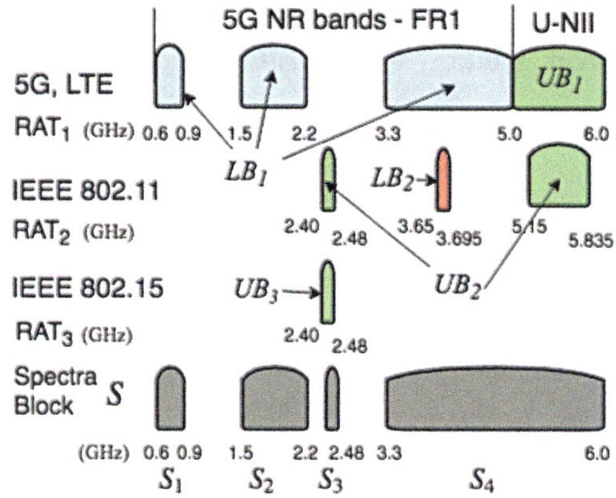

Fig. 5.1 An example of the spectra block $\mathcal{S}$ in sub-6GHz frequency bands

5.2.1 System Model

Let time t proceed in discrete steps, i.e., $t \in \mathcal{T} = \{1, 2, \cdots\}$. Consider a DSA-enabled system deployed in a geographical region $\mathcal{A} \subset \mathbb{R}^2$. The spectrum of interest, $\mathcal{S}$, refers to the spectrum blocks that are shared[1] among the K radio access technology (RAT) $\{\text{RAT}_i\}_{i=1}^K$ allowed in this system.

5.2.1.1 Spectra of Interest $\mathcal{S}$

Each RAT_i has a licensed band LB_i exclusively reserved for authorized RAT_i users, and an unlicensed band UB_i to be shared with users accessing via other RAT's. Each LB_i or UB_i can be viewed as an *interval* identified by the lowest and highest frequency as its endpoints (or a union of such intervals), then the union of all licensed and unlicensed bands, $\mathcal{S} := \cup_{i=1}^K \{\text{LB}_i \cup \text{UB}_i\}$, is the spectral block that a SAS process takes place in a DSA-enabled system.

An example of the spectral block is shown in Fig. 5.1, where $K = 3$ RAT's are allowed in the system: cellular (Long Term Evolution (LTE)/5G, RAT_1), IEEE 802.11 (WiFi, RAT_2), and IEEE 802.15 (Bluetooth, RAT_3). Among these, RAT_1 has the licensed 5G New Radio (5G NR) FR1 bands [1] to itself, as indicated by LB_1, while its unlicensed U-NII bands UB_1 are shared with RAT_2, on which licensed-assisted LTE access co-exists with WiFi access [8, 32]. Meanwhile, the unlicensed ISM bands UB_3 are shared by RAT_2 and RAT_3. The spectra of interest $\mathcal{S} = \cup_{i=1}^3 S_i$ is their union.

[1] There are two spectrum-sharing scopes for a DSA-enabled system: the inter-technology DSA, which only shares the unlicensed spectrum bands [8], and the *spectrum commons*, in which licensed bands are also included, and each device has equal spectrum access right on a cost basis [32]. Both scopes can be described by our model.

Without loss of generality,[2] we write $\mathcal{S}$ as interval $[s_L, s_H] \subset \mathbb{R}$, and further divide it into $\lceil \frac{s_H - s_L}{\Delta f} \rceil$ *spectrum slices* of width Δf, which is determined by:

1. *Channel bandwidth* of $\{\text{RAT}_i\}_{i=1}^K$. There may not be a unified channel access scheme on $\mathcal{S}$ when $K > 1$. For instance, the U-NII bands can be accessed through LTE and WiFi. For the former, the standard channel bandwidth are 1.4, 3, 5, 10, 15, and 20 MHz, while for the latter, the channel bandwidth ranges from 10 to 160 MHz. Further, an LTE channel is divided into resource blocks (180 KHz) that contain 12 sub-carriers, while each IEEE 802.11n channel (20 MHz) contains 52 sub-carriers of 312.5 KHz. Therefore, we choose the slice width Δf as a common divisor of all the channel bandwidths allowed by the K RAT's, such that a channel under each RAT_i contains k_i spectrum slices, where $k_i \in \mathbb{N}^+$.
2. *Resolution bandwidth* of monitoring devices. Due to the different sampling rates of commercial/prototype monitoring hardware, e.g., 10 MS/s for USRP E310, and 2.4 MS/s for the low-cost SDR prototype designed in [25], which are constrained by their processing power (especially for FFT), and the stabilizing time of the sweep-tune process, the resolution bandwidth of spectrum monitors are subject to various limits. Typically, it is set to be 1–3% of the channel bandwidth [15, 26] for observable results, but it is also required to be greater than 1 KHz to avoid overloading the processors [26].

Based on these facts, a spectrum slice of width Δf is used as the smallest[3] unit of spectrum trunk to be associated with an *access specification* and a time-varying observable *state*, defined as follows.

For each spectrum slice, an *access specification* states the legitimate way to access this slice, including allowed radio access technology (RAT), maximum transmitting power, maximum aggregated channel bandwidth, register/authentication procedure, and so on. For example, an 1 KHz slice in the 5G NR FR1 (LB_1 in Fig. 5.1) can only be accessed through LTE/5G, with transmission settings specified in 3GPP technical specifications, e.g., [1]. In this way, spectra block S is a database with $\lceil \frac{s_H - s_L}{\Delta f} \rceil$ items, against which monitors checks activities on each slice.

The *state* of a spectrum slice $i \in [1, \lceil \frac{s_H - s_L}{\Delta f} \rceil]$ (frequency range $[s_L + (i-1)\Delta f, s_L + i\Delta f)$) is the result of spectrum activities on this particular slice. Based on the activities, spectrum slice i can be in either of the three states:

- *Idle*, when it is not occupied by any user in the current time step, e.g., the white slice $i = 4$ and slice $i = 5$ in time step $t = 1$ in Fig. 5.2.

[2] Spectra block $\mathcal{S}$ in an wireless overlay system may not form one single continuous interval, rather, it is the union of several non-overlapping continuous intervals, i.e., $\mathcal{S} = \mathcal{S}_1 \cup \mathcal{S}_2 \cup \cdots$. We focus on one of those intervals in this chapter, for the simplicity of notation and understanding.

[3] Note that Δf is not the frequency range that can be scanned by a monitor during a time step. For example, in Fig. 5.2 a monitor (red/blue box) can determine the states of 4 spectrum slices.

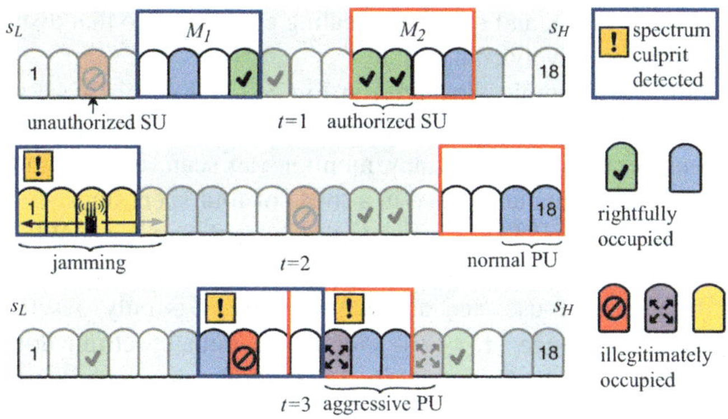

Fig. 5.2 Two spectrum monitors M_1 and M_2 watch over spectrum block $\mathcal{S} = [s_L, s_H]$. Each slice can be in *idle* (in white) or *occupied* (colored) state, due to various spectrum activities

- *Rightfully occupied*, if it is accessed by a device obeying the access specification of this slice. For instance, the blue slices {17, 18} are rightfully occupied by an authenticated primary user (PU) at time $t = 2$; the green slice 8 and slices {12, 13} (with ✔ marker) are accessed by authorized secondary user (SU) at time $t = 1$.
- *Illegitimately occupied*, if the occupant does not comply with the access specification of slice i. For instance, the purple slices {11, 14} are beyond the designated bands of the aggressive PU at time $t = 3$; the red slices (with restriction sign) are used by an unauthorized SU during $t = 1$ to $t = 3$; the yellow slices [1, 6] are jammed by an attacker emitting a high-power signal at $t = 2$. We refer to these illegitimate occupants as *spectrum culprits*, to be detected by monitors.

Rightful or illegitimate, all aforementioned spectrum activities take place in a 3-D space that spans over both the spectrum domain and the geographical domain.

5.2.1.2 The Spectra-Location Space X

Consider the monitoring process of 1-D spectrum $\mathcal{S}$ over a closed 2-D geographical region $\mathcal{A} \subset \mathbb{R}^2$. Together, they compose a 3-D product space $\mathcal{S} \times \mathcal{A}$, referred to as the *spectra-location space* X. Then for any point $x \in X$, there exist projection maps $p_A : X \to \mathcal{A}$ and $p_S : X \to \mathcal{S}$ that identify the frequency and location of any point $x \in X$ respectively. In the product space X, the *spectra-location distance* d_{SA} between point x_i and point $x_j \in X$ is defined as the product metric induced by the Euclidean distance metrics in the spectrum domain $\mathcal{S}$ and the space domain $\mathcal{A}$, i.e.,

$$d_{SA}(x_i, x_j) := \|d_S\left(p_S(x_i), p_S(x_j)\right), \frac{1}{\epsilon} d_A\left(p_A(x_i), p_A(x_j)\right)\|_2, \tag{5.1}$$

where $\|\cdot\|_p$ is the p norm, and $\epsilon > 0$ is a scaling coefficient so that distance d_A and distance d_s are quantitatively comparable. [4]

In this sense, the example illustrated in Fig. 5.2 is a special case when region $\mathcal{A}$ shrinks to a point $\{a\}$, such that merely the spectrum domain $\mathcal{S}$ needs to be taken into consideration. But still, any monitor can scan, examine and/or record the spectrum activities that take place in a 'box' of four spectrum slices, as a result of which the unauthorized SU is not detected until time $t = 3$. When the space domain $\mathcal{A}$ is sufficiently large, spectrum slices are annotated with locations, due to possible frequency re-use, and a spectrum activity is only observable within the sensing/detection range [15] of a monitor in both spectrum domain $\mathcal{S}$ and geographical space domain $\mathcal{A}$, as described in the following surveillance model.

5.2.1.3 Surveillance Model

Denote $\mathcal{M} = \{M_1, M_2, \cdots, M_m\}$ as the set of m monitors in the system. For a single monitor, it can only determine the states of adjacent slices [15] during a time step, as illustrated by the boxes in Fig. 5.2. Moreover, due to the attenuation of wireless signal over distance, such constraint also exists in the space domain $\mathcal{A}$. As a result, the capability of a monitor to observe spectrum activities is restricted to a closed space, defined as the *monitoring power*.

Definition 5.1 For a monitor $M \in \mathcal{M}$ assigned at location $a^t_M \in \mathcal{A}$ and center frequency $s^t_M \in \mathcal{S}$ at time t, the **monitoring power** of monitor M is defined as a δ-ball centered at $s^t_M \times a^t_M \in X$, that is,

$$Ball_\delta(s^t_M \times a^t_M) := \{x \in X | d_{SA}(s^t_M \times a^t_M, x) \leq \delta\}, \tag{5.2}$$

inside which spectrum activities (rightful/illegitimate occupancy) can be identified by monitor M with probability q.

In this definition, parameter δ and q capture the *locality* and the probabilistic nature of spectrum monitoring activities, respectively. For a spectrum monitor, these two parameters are determined by hardware performances, including sensitivity, noise floor, input range, etc. and the detection algorithm. By fine-tuning function $q(\cdot)$, radius δ, and parameter ϵ in Eq. (5.1), a variety of monitoring techniques can be depicted by this δ-ball model.

The *closed* δ-ball[5] describes the surveillance range of a single monitor, as illustrated in Fig. 5.3(left). Parameter δ captures the *locality* of surveillance: the

[4] By comparable, we mean their impacts on monitoring strategy design, e.g., easiness to relocate, energy or time consumption, are on the same order in quantity.

[5] Technically, some spectrum slices will be divided by the sphere, excluding which the monitoring power is not a smooth ball. We still refer to the monitoring power as δ-ball for the ease of comprehension. Spectrum slices being divided is not a problem, because: (1) the slice width Δ_f is oftentimes much thinner than the width $s_H - s_L$ of the spectra block $\mathcal{S}$ (e.g., the 775 MHz U-NII

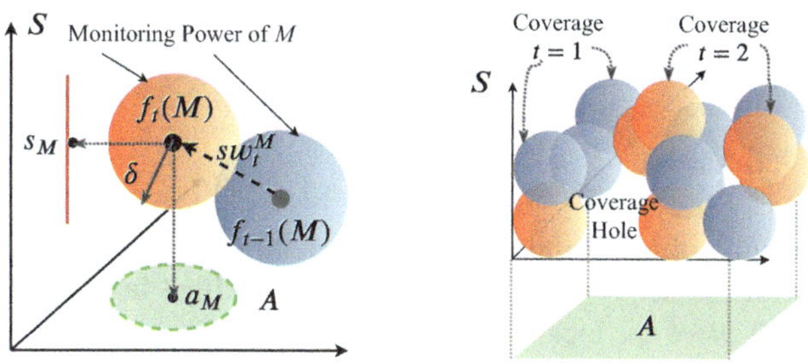

Fig. 5.3 The $q(\delta)$-monitoring power, switching action sw_t^M, and the strategy coverage in spectra-location space $X = \mathcal{S} \times \mathcal{A}$

monitoring power of a single monitor is the result of the trade-off between spectrum range (in spectrum distance d_S) and geographical range (in distance d_A), due to the limited sampling rate imposed by hard-ware constraints [25, 35]. To be more specific, the number of samples that a monitor can collect per unit time is *limited* [15], and these samples can be employed to cover either a larger bandwidth (large d_S) with a lower sensitivity, or a narrower bandwidth with a higher sensitivity. In other words, large d_S and large d_A can not be achieved simultaneously by a single monitor, hence the closed ball shape. For the most commonly-used energy detection method, e.g., in [20, 25], lower sensitivity translates to a higher power threshold, resulting in a reduced detecting range in geographical domain, i.e., small d_A.

Function $q : \mathbb{R} \to [0, 1]$ quantifies the reliability of results within the monitoring power, or equivalently the detection probability[6] of spectrum culprits. It has the following properties: (1) q is a surjective; (2) q is non-decreasing in $\mathbb{R}$; (3) we can define its inversion $q^{-1} : [0, 1] \to \mathbb{R}$ as $q^{-1}(y) := \sup_{x>0}\{q(x) = y\}$, so for any required reliability $y \in [0, 1]$, there exists a critical radius $\delta^* = q^{-1}(y)$, above which the monitoring results are not acceptable. Consequently, if a point $x \in X$ is covered by the $q(\delta)$-monitoring power of k monitors, illegitimate occupancy at this point x can be detected with a higher probability, that is, $1 - [1 - q(\delta)]^k$.

Remark 5.1 By fine-tuning function $q(\cdot)$, radius δ, and parameter ϵ in Eq. (5.1), a variety of monitoring techniques can be depicted by this δ-ball model. These parameters are determined by hardware performances, including sensitivity, noise floor, input range, and the detection algorithm. Radius δ refers to $\delta_* = q^{-1}(1)$ that

frequency bands); (2) the δ-balls are forced to overlap in the space-tessellation step, such that there will not be coverage gaps.

[6] For any radius δ, it is more probable to determine whether a spectrum slice at a location is occupied or not ($q_c(\delta)$), than determining whether the occupancy is legit ($q_c(\delta)$). Consequently, $q_c(\delta)$ for occupancy measurement is greater than or equal to that for culprit detection $q_d(\delta)$. We set $\delta = \max\{\delta > 0 \mid q_c(\delta) = 1\}$, and $q = q_d(\delta)$, such that occupancy measurement is accurate, while culprit detection is not.

can guarantee a fully reliable detection result, if not explicitly specified hereafter, in the remaining of this chapter.

5.2.1.4 Exploit Model

Recall a spectrum culprit at time t is the occupant of a spectrum slice that does not comply with the access specifications, as exemplified in Fig. 5.2. Consider a spectrum culprit $R \in \mathcal{R}$ located at $a_R \in \mathcal{A}$, who illegitimately occupies one or multiple spectrum slices, denoted as $S_R \subset \mathcal{S}$, at time t. Culprit R leaves a 'mark' $R_t = S_R \times a_R \subset X$. The wider S_R is, the larger the 'mark', and the more detectable R becomes. For analysis reasons, we consider spectrum culprits that are most difficult to detect as the worst-case scenario, that is, S_R shrinks to a point $\{s_R\} \in \mathcal{S}$, such that we can write $R_t \in X$; we also assume that, throughout time $\mathcal{T}$, culprits stays in the system and continues its spectrum exploit.[7]

Then over time $\mathcal{T}$, the exploit marks of culprit R constitute an *exploit sequence* $\{R_t\}_{t\in\mathcal{T}}$. The detailed exploit *pattern* of spectrum culprits, i.e., how a spectrum culprit $R \in \mathcal{R}$ assigns its exploit sequence, can be either oblivious or adversarial, depending on its learning capability, and will be discussed later in details. At any time t, culprit R is considered to be *detectable* with probability at least $q(\delta)$, if the exploit mark R_t overlaps with the monitoring power of some monitors, i.e., $\exists M_i \in \mathcal{M}$ such that $R_t \cap Ball_\delta(f_t(M_i)) \neq \phi$.

5.2.1.5 Switching Model

Despite their different roles in the DSA-enabled system, normal users (denoted as $\mathcal{U}$), spectrum culprits ($\mathcal{R}$), and monitors ($\mathcal{M}$), are all wireless devices capable of moving and tuning. Both actions will result in a relocation of devices in the spectra-location space X, which we refer to as a *switching* sw_t^Y of device $Y \in \mathcal{U} \cup \mathcal{M} \cup \mathcal{R}$, that is, a move of Y from point $Y_{t-1} \in X$ to point $Y_t \in X$. For example, in Fig. 5.3(left), the switching sw_t^M corresponds to the relocation and tuning of monitor M between $t-1$ and t, hence the change of its monitoring power (from the blue ball to the red ball). As opposed to the assumption in [18, 25], this common switching action is also constrained in X, due to the induced *cost* in the form of time, energy, budget, and so on, which is a design concern for SAS process in some

[7] In reality, it is possible that a culprit occasionally misuse some spectrum slices, or even leave the system indefinitely after such exploit. To quantify the performance of a SAS system based on its ability to catch these culprits would not be objective, or fair, as it depends on the culprits' spontaneous behavior. Therefore, we assume culprits stays in the system across $\mathcal{T}$. On the other hand, whenever a culprit's misconduct is recorded for the first time, some attributes, e.g., spectrum fingerprint [4], can be subtracted and distributed to all monitors, such that a match can be found even if the culprit has been accessing the spectrum legitimately ever since.

scenarios. Therefore, we define the *switching capacity* to capture this constraint in space X.

Definition 5.2 Let $Y_t \in X$ denote the location (in space X) of device Y at time t, the **Switching Capacity** α_Y of Y is defined as the maximum distance in X, that device Y can switch over by one action in a time step, that is,

$$\alpha_Y := \sup_{t \in \mathcal{T}} \{d_{SA}(Y_t, Y_{t+1})\}. \tag{5.3}$$

Device Y is referred to as an α_Y-monitor or α_Y-culprit.

The switching capacity limit can take different values in different SAS scenarios. For a dedicated (physical) monitor, such as specialty monitoring hardware mounted on a drone or vehicle, a switching action is composed of physical movement and/or tuning. Consequently, strategy design is restricted by a quantitative switching cost, including time, energy, budget etc., which is a function of the switching distances $d_{SA}(\cdot, \cdot)$. On the contrary, switching actions in a crowdsourcing scenario are merely changes of surrogate (logical) monitors, which are wireless devices that are capable of and willing to monitor the spectrum activities for the system. Consequently, if immediate communication among participants is guaranteed, or there exists a central controller capable of timely coordination, switching will not be constrained, i.e., $\alpha_M = \infty$; otherwise for the case of distributed control, which relies on local wireless communication, switching may not possible beyond the communication range of participating devices. Designing SAS strategies for these two scenarios are discussed in Sects. 5.4 and 5.5, respectively.

5.2.2 *Performance Metrics*

To formulate the SAS strategy design problem, we first need to formally design a *strategy* with respect to the m monitors. At the beginning of time step t, each monitor $M_i \in \mathcal{M}$ is assigned to a spectra-location point $f_t^m(M_i) \in X$, through an *assignment* map $f_t^m : \mathcal{M} \to X^m$, e.g., f_1 and f_2 in Fig. 5.3 (left). Allowing time t to proceed in $\mathcal{T}$, assignment points of the m monitors constitute a *strategy*.

Definition 5.3 (Strategy[8] f_t^m) A strategy $\{f_t^m\}_{t \le T}$ is a sequence of assignments during time interval $[1, T] \subset \mathcal{T}$, subject to the switching capacity constraints $\{\alpha_{M_i}\}_{M_i \in \mathcal{M}}$. During time step t, monitors in set $\mathcal{M}$ can scan

[8] Superscript m in f_t^m denotes the number of monitors, while subscript t denotes time. A second subscript may be added to differentiate strategy types, e.g., $f_{S,t}^m$ for deterministic strategy. Any of the three denotations (number of monitors, time and type) may be omitted, when no confusion is raised.

$$C(f_t^m) = \bigcup_{M_i \in M} Ball_\delta\left(f_t^m(M_i)\right) \tag{5.4}$$

en masse, which is referred to the (surveillance) coverage of assignment f_t^m. Sweep-coverage of a strategy $C(f)$ is then the union of sequence $\{C(f_t^m)\}_{t\in[1,T]}$.

Recall the two objectives of a SAS process, that is, spectrum occupancy measurement and culprit detection. The former urges for a quick sweep-scan of the entire spectra-location space X, i.e., minimizing the time needed to satisfy the coverage goal $X \subset C(f)$, such that spectrum (occupancy) status can be timely recorded and updated to users. The latter requires effective detection of spectrum culprits, such that the time that an undetected culprit illegitimately occupies spectrum slices can be reduced. For instance, in the special case scenario ($X = \mathcal{S} \times \{a\}$) illustrated in Fig. 5.2, the entire X is sweep-covered at $t = 3$, and the unauthorized SU (culprit) exploited the system for two time steps before its detection at $t = 3$. In other words, the efficacy of a strategy can be quantitatively evaluated and fairly compared through the following two metrics, with respective to the coverage and detection goals.

Definition 5.4 Under strategy $\{f_t^m\}_{t\in T}$, the **coverage time** T_f^m is defined as the first time that its sweep-coverage $C_T(f^m)$ contains every point in space $X = \mathcal{S} \times \mathcal{A}$, that is,

$$T_f^m := \min\{T \in \mathcal{T} \mid x \in C_T(f^m),\ \forall x \in X\}. \tag{5.5}$$

The **detection time** $\tau_R(f^m)$ of a culprit R with exploit sequence $\{R(t)\}_{t\in T}$, is defined as the first time that culprit R can be identified by any of the m monitors, that is,

$$\tau_R(f^m) := \min\{t \in \mathcal{T} \mid \left[\sum_{i=1}^{m} \mathbb{1}_{R_t \in Ball_\delta(f_t^m(M_i))} D_i\right] \geq 1\}, \tag{5.6}$$

where detection outcome D_i is a Bernoulli r.v. with mean q.

Particularly for $\delta = \delta_*$, when monitoring (detection) result is fully reliable, i.e. $q(\delta_*) = 1$, the detection time can be further simplified to

$$\tau_R(f^m) := \min\{t \geq 1 \mid R(t) \in C(f_t^m)\}. \tag{5.7}$$

5.2.3 SAS Strategy Design Problem

With the system model described above, this chapter studies the SAS process for a set of m α_M-monitors to achieve the sweep-coverage and culprit detection goals in the spectra-location space X. Specifically, we intend to design strategies $\{f_t^m\}_{t\in\mathcal{T}} \in \{X^m\}^{\mathcal{T}}$ for dedicated and crowdsourcing SAS scenarios respectively, and examine their *efficacy* by answering the following questions:

- What is the coverage time T_f of the designed strategy f^m, by which time spectra-location space X is sweep-covered, i.e., $X \subset C(f^m)$?
- Under strategy f, what is the detection time $\tau_R(f^m)$ of a spectrum culprit $R \in \mathcal{R}$ with the exploit sequence $\{R_t\}_{t\in[1,\mathcal{T}]}$?

5.3 A Two-Step Solution

Designing an SAS strategy $\{f_t^m\}_{t\in\mathcal{T}} \in \{X^m\}^{\mathcal{T}}$ is equivalent to finding a sequence of assignment points in the spectra-location space X for every monitor $M \in \mathcal{M}$, at every time instant t, subjecting to the switching capacity constraint α_M. There are two major challenges in this process: First, for every time t, the solution space X^m is of infinite size, which hinders both the analysis-based approach and the search-based experiment approach. Second, switching actions of monitors, i.e., the tuning (a move in the spectrum domain $\mathcal{S}$) and/or relocation (a move in the geographical space domain $\mathcal{A}$), can also be constrained by the accompanied cost.

To overcome these challenges, we propose a two-step solution: first, the continuous strategy space $\{X^m\}^{\mathcal{T}}$ is reduced to a discrete and *finite* space $\{V^m\}^{\mathcal{T}}$ through space-tessellation; then any surveillance strategy is formulated as a *walk* on the graph, whose edges illustrate possible switching actions of monitors. In this way, SAS as a global activity is transformed into a chain of individual actions, i.e., switching (walking) of monitors and culprits, such that design of SAS strategies becomes tractable in both dedicated and crowdsourcing scenarios.

5.3.1 Space Tessellation: Reducing the Solution Space

Driven by the sweep-coverage and culprit detection objectives, the assignment points of monitors $\{f_t(\mathcal{M})\}_{t\in[1,T_f]}$ of a good SAS strategy should have the following properties:

1. *Least points*. To timely update the spectrum occupancy data, monitors are expected to sweep-scan the entire spectra-location space X as quickly as possible, which translate to achieving the coverage goal $X \subset \cup_{t=1}^{T_f} C(f_t)$ with as few (mT_f) assignment points as possible.
2. *Minimal overlapping*. To quickly detect culprits, every assignment f_t is expected to cover as much non-overlapping space (large $C(f_t)$) as possible, which translates to a minimal overlapping of monitoring powers during every assignment.[9]

[9] Though overlapping monitoring power permits a higher detection probability inside the δ-ball than that of a single one, it is not necessary when the detection probability q is sufficiently high.

These requirements can be jointly satisfied if the continuous space X is divided into a minimum number of non-overlapping *cells*, each covered/contained by a δ-ball, i.e. the monitoring power of a monitor, and every assignment map f_t takes value from the set V of cell centers, instead of the entire X. In this way, the first step becomes a *tessellation* problem of space X.

5.3.1.1 Solution to the Space-Tessellation Problem

Space tessellation, or honeycomb, in the 3-D space X, refers to the close packing of 3-D *cells* without overlaps or gaps. So the objective of the tessellation problem is to find the best form of cells, which can jointly cover a fixed X with the least cells.

A cell in the tessellation can be regarded as the non-overlapping part of a δ-ball (monitoring power), and comes in various forms in the solution to this classic problem, e.g., cube, hexagon-prism, tetrahedra, etc. The higher volume-efficiency ρ of a cell, that is, the volume ratio of the cell over its in subscribed δ-ball, the more efficient the form of the cell, due to the less overlap between adjacent monitoring power (δ-balls). It would be ideal to fully utilize δ-balls to fill the space. However, direct packing of solid balls always leaves gaps, i.e., spectrum holes in the sweep-coverage, which can be eliminated by pushing the 'elastic' balls into each other. Meanwhile each 'squeezed' ball (the resulting cells) should be inscribed to a δ-ball, i.e. the maximum distance of any two points on the cell surface is required to be smaller than 2δ.

The space tessellation problem is closely related to the *Kelvin problem* [31], which aims to find the most efficient bubble wrap form to fill a space with the least surface area. In the Kelvin problem, efficiency is quantified by the *isoperimetric quotient* [33], and the ball shape (as the monitor power in our model) has the highest isoperimetric quotient value of 1. Therefore, the most efficient cell form in the tessellation problem is the one that has the highest isoperimetric quotient. In this sense, the best solution for wide (large $|\mathcal{S}|$) spectrum is the *Kelvin structure* (with isoperimetic quotient value 0.757), whose cells are truncated octahedrons (Fig. 5.4), arranged in a layered manner (Fig. 5.5). The truncated octahedron cell has the highest volume-efficiency ρ compared to other forms, as shown in Table 5.1. Centers of these cells correspond to assignment points composing the discrete and *finite assignment space* V, whose size $n = |V|$ can be obtained from the following proposition 5.1.

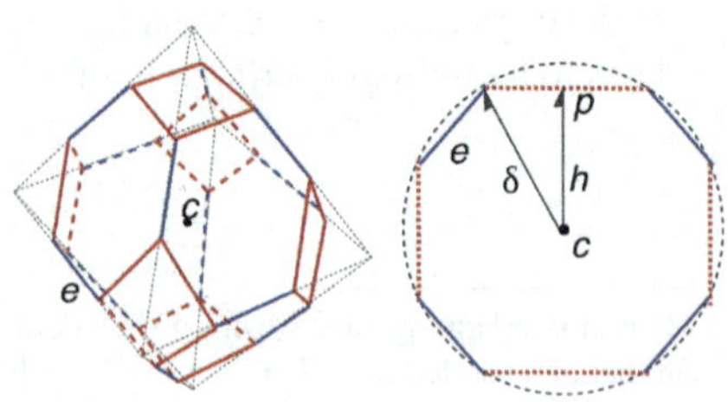

Fig. 5.4 The most efficient cell is truncated octahedron, whose top view is also shown

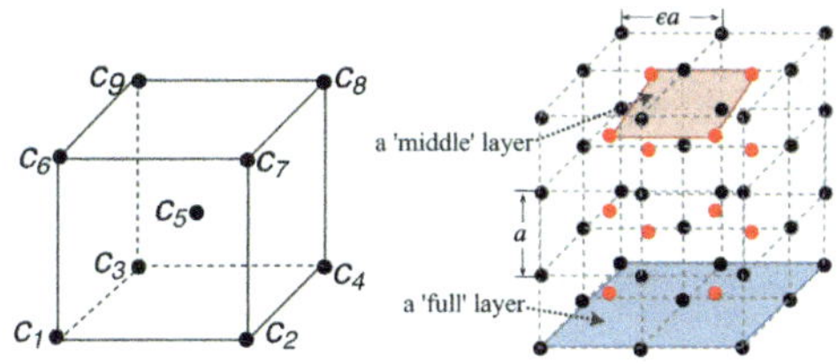

Fig. 5.5 To pack these cells densely, their centers $\{c_i\}_i$ will be arranged into alternating layers. So the discrete assignment space V is composed of cell centers of the Kelvin structure, in the form of 'full' (black) and 'middle' (red) layers

Table 5.1 Volume efficiency of different cell forms discussed in this chapter

Cell form	Volume efficiency ρ	Isoperimetric quotient[a]	Size $n = \|V\|$
Cube	$\geq \frac{2}{\sqrt{3}\pi} \simeq 0.368$	$\frac{1}{216}$	$\lceil \frac{\sqrt{3}\|S\|}{2\delta} \rceil \cdot \lceil \frac{3\|\mathcal{A}\|}{8(\epsilon\delta)^2} \rceil$
Hexagon-prism	$\geq \frac{3\sqrt{3}}{2}\alpha\sqrt{(1-\alpha^2)} \leq 0.585$	$\frac{3\sqrt{3}b}{2(3\sqrt{3}+12b)^3}$	$\frac{2\|\mathcal{A}\|}{3\sqrt{3}(1-\alpha^2)(\epsilon\delta)^2} \rceil$
Truncated octahedron	$\geq \frac{24}{5\sqrt{5}\pi} \simeq 0.683$	0.757	$\lceil \frac{5\sqrt{5}\|S\|\|\mathcal{A}\|}{16\epsilon^2\delta^3} \rceil$

[a]Constant $b = \sqrt{\frac{|S|^2}{4\delta^2 - |S|^2}}$ for the hexagon-prism cell

Proposition 5.1 *When the spectrum block $\mathcal{S}$ is narrow ($|\mathcal{S}| = 2\alpha\delta$, $0 < \alpha << 1$), the size of the assignment space $n = |V|$ can be determined by tessellation with hexagon-prism cells, i.e.,*

$$n_{hex} = \lceil \frac{2A}{3\sqrt{3}(1-\alpha^2)(\epsilon\delta)^2} \rceil, \tag{5.8}$$

where A denotes the area of region $\mathcal{A}$, δ corresponds to the monitoring power and ϵ is the scaling coefficient in Eq. (5.1). *Otherwise, size n is achieved by tessellation with truncated octahedron (Kelvin structure) cells, and*

$$n_o \geq \lceil \frac{5\sqrt{5}|\mathcal{S}|A}{16\epsilon^2\delta^3} \rceil. \tag{5.9}$$

Further, if the geographical region $\mathcal{A}$ is rectangular,

$$n_o \geq \lceil \frac{\sqrt{5}|\mathcal{S}|}{4\delta} \rceil \cdot \lceil \frac{5A + 2\sqrt{5A}\delta(3-2\epsilon) + 4\delta^2(1-\epsilon)^2}{8(\epsilon\delta)^2} \rceil, \tag{5.10}$$

where $e = \frac{2}{\sqrt{10}}\delta$ is the edge length of cells.

Proof As shown in Fig. 5.6, the narrow spectrum block, i.e., $|\mathcal{S}| = 2\alpha\delta$, implies that one layer of δ-balls is enough to cover the entire spectrum $\mathcal{S}$, if all of the monitor

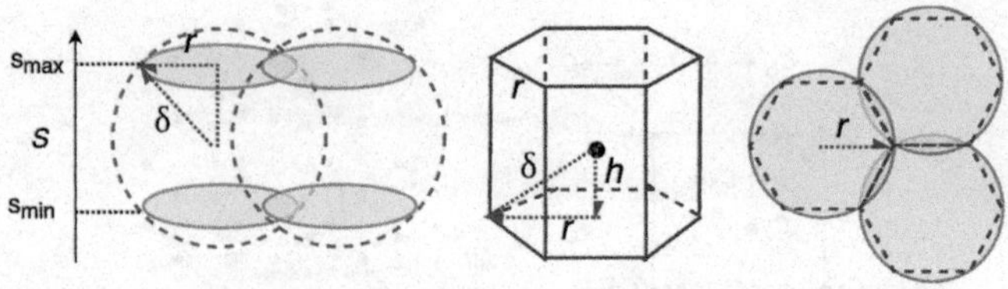

Fig. 5.6 Tessellation of X with hexagonal prism when spectrum range is narrow

radios are tuned to the center frequency $s = \frac{s_{max}+s_{min}}{2}$. At s_L and s_H, each δ-ball creates a disk print with radius r on the surface of X, as shown in blue shade in Fig. 5.6. Then the 3-D tessellation problem becomes the coverage of a planar area $\mathcal{A}$ with disks of radius $r = \sqrt{(1-\alpha^2)}\delta$, as illustrated in Fig. 5.6 (right). The effective non-overlapping part (a cell) is a hexagonal prism.

To cover area $\mathcal{A}$ with the least overlap among disks, the disks should be arranged according to the hexagonal tessellation, with each hexagon inscribed to a circle, as in Fig. 5.6 (right). The edge length of any hexagon $r = \sqrt{(1-\alpha^2)}\delta$. Bound on the number of assignment points n_{hex} in Eq. (5.8) follows from hexagonal geometry. Also, the coverage of the boundary points (the top and bottom surface) are guaranteed.

For broader spectrum $\mathcal{S}$, the best known 3-D tessellation form is the Weaire-Phelan structure[33], which has two types of cells with the same isoperimetic quotient value 0.764. However, the arrangement of two types of cells makes it more difficult to design monitor assignment strategies. Therefore, we adopt the sub-optimal *Kelvin structure*, whose cell is a truncated octahedron with isoperimetic quotient value 0.757,[10] as shown in Fig. 5.4 (left). Each cell is a polyhedron with six square faces and eight hexagonal faces, both with edge length e. The center of each cell coincide with the center c of its outscribed δ-ball, so its largest cross section through the center c is an octagon with edge length $p\sqrt{2}e$.

Since $e = \frac{2}{\sqrt{10}}\delta$, the volume efficiency $\rho_{oct} = \frac{24}{5\sqrt{5}\pi} \simeq 0.683$, much higher than the hexagon prism cell and the cubic cell, as listed in Table 5.1. To form a 3-D tessellation, the arrangement pattern of cells is body-centered cubic, as shown in Fig. 5.5. Then lower bound of n in Eq.(5.9) can be obtained by

$$n_o \geq \frac{|\mathcal{S}||\mathcal{A}|}{\epsilon^2 \mathrm{Vo}^o} = \lceil \frac{5\sqrt{5}|\mathcal{S}|A}{16\epsilon^2\delta^3} \rceil.$$

A tighter bound can be found if $\mathcal{A}$ is a rectangular region. Suppose the length of $\mathcal{A}$ is l, and the width is w. The number of 'full' layers (blue shade in Fig. 5.5 right) is

[10] The Kelvin structure is allowed to have curved surfaces to have a smaller surface area, which is not necessary in our case.

$$L_{full} = \lceil \frac{|S|}{2h} + 1 \rceil = \lceil \frac{\sqrt{5}|S|}{4\delta} \rceil + 1, \tag{5.11}$$

and the number of "middle" layers (red shade in Fig. 5.5 right) is $L_{mid} = L_{full} - 1$. For the coverage of region $\mathcal{A}$, the number of cells in a "full" layer can be lower bounded by

$$N_{full} = \lceil \frac{l + \sqrt{2}e}{2h \cdot \epsilon} \rceil \cdot \lceil \frac{w + \sqrt{2}e}{2h \cdot \epsilon} \rceil \geq \lceil \frac{5A + 4\sqrt{5A}\delta + 4\delta^2}{16(\epsilon\delta)^2} \rceil. \tag{5.12}$$

And the number of cells in a middle layers can be lower bounded by

$$\begin{aligned} N_{mid} &= \lceil \frac{l + \sqrt{2}e}{2h \cdot \epsilon} - 1 \rceil \cdot \lceil \frac{w + \sqrt{2}e}{2h \cdot \epsilon} - 1 \rceil \\ &\geq \lceil \frac{5A + 8(1-\epsilon)\sqrt{5A}\delta + 4\delta^2(1-4\epsilon)}{16(\epsilon\delta)^2} \rceil + 1. \end{aligned} \tag{5.13}$$

Then the total number of truncated octahedron cells needed is

$$n_o = L_{full} N_{full} + L_{mid} N_{mid}, \tag{5.14}$$

which yields the tighter bound Eq.(5.10) with some manipulations. □

By space-tessellation, sweep-coverage of X is guaranteed as long as the assignment maps $\{f_t\}_t$ are jointly surjective on the *finite* assignment space V of cell centers. In other words, it is sufficient for monitors ($\mathcal{M}$) to switch among assignment points in V, which greatly reduces the size of the solution space (from ∞ to $|V|^m$) for every time step. For the ease of notation and discussion, we can also restrict the range of exploit points to V, because any point x in X can be uniquely mapped to a cell with its center v_x in V, and the probability of a spectrum culprit $R \in \mathcal{R}$ being detected at $R_t \in X$ is the same as that at $R_t = v_x$.

5.3.1.2 Exploit Patterns of Culprits in Assignment Space V

After the preparation of setting up the assignment space V, we categorize the exploit patterns of spectrum culprits in this context, i.e., how a spectrum culprit $R \in \mathcal{R}$ determines its exploit point $R_t \in V$ for the next time step, with respect to its learning capabilities.

Definition 5.5 A **persistent culprit** $R_p \in \mathcal{R}$ refers to a spectrum culprit, whose exploit sequence $\{R_t\}_t$ does not change over time, that is, $\{R_t\}_t$ is composed of i.i.d. r.v.'s R_t^p, all distributed with PMF $g_{R_p}(v)$, where v is an assignment point in V.

Table 5.2 Three types of persistent spectrum culprits

Type	Culprit description	PMF $g_{R_*}(\cdot)$	Detection time[a]
R_s	Stationary, fix-frequency	$g_{R_s}(v) = \mathbf{1}_{v_s}(v),\ \forall v \in V$	$\mathbb{E}(\tau_s(f_S)) = \frac{T_s}{2}$
R_{sd}	Stationary, DSA-enabled	$g_{R_s}(v) = \begin{cases} \frac{1}{\lvert V_{sd}\rvert}, & \text{if } v \in V_{sd}, \\ 0, & \text{otherwise.} \end{cases}$	$\mathbb{E}(\tau_{sd}(f_S)) \leq \frac{n}{m} \simeq T_s$
R_{md}	Mobile, DSA-enabled	$g_{R_{md}}(v) = \frac{1}{n},\ \forall v \in V$	$\mathbb{E}(\tau_{md}(f_S)) = \frac{n}{m} \simeq T_s$

[a]The expected detection time of the persistent culprits, including quantity T_s, are discussed in Sect. 5.4.2.1

Persistent culprit R_p can describe a variety of exploiting strategies with different PMF $g_{R_p}(v)$. For instance, as shown in Table 5.2, R_s is a stationary culprit that can only access a selective range of frequencies; R_{sd} corresponds to a DSA-enabled stationary culprit; and R_{md} is a mobile DSA-enabled culprit that moves in region $\mathcal{A}$.

As opposed to oblivious persistent culprits, machine learning-assisted RAT [24, 30] allows sophisticated culprits to steer the game toward their benefit, by actively dodging monitors [18]. The intuition behind adversarial culprits is that, once a SAS strategy is known[11] by a culprit R_a, R_a can then switch to points that are less probable to be monitored in the next time step.

Definition 5.6 An **adversarial culprit** $R_a \in \mathcal{R}$ is a spectrum culprit with prior knowledge of current strategy $\{f_t^m\}_{t\in\mathcal{T}}$, that is, R_a knows the set of probabilities $\{v \in \cup_{M_i} f_t^m(M_i)\}_{v\in V}$ ahead of time t, and determines its current exploit point R_t^a with PMF $g_{R_a}^t(v) = \frac{1}{|\mathrm{Void}(t)|}$ for point $v \in \mathrm{Void}(t)$, where $\mathrm{Void}(t) = \arg\min_{v\in V} \mathbb{P}\Big(v \in \bigcup_{M_i\in\mathcal{M}} f_t(M_i)\Big)$ is the 'spectrum hole' that is least likely to be monitored at t.

Note there is no switching constraint in Definitions 5.5 and 5.6, which describe the most powerful culprits in terms of switching, i.e., $\alpha_R = \infty$. As discussed before for monitors, it is possible that culprit R has $\alpha_R < \infty$, such that the range of R_{t+1} will be restricted to a smaller subset $N(R_t) = \{v \in V \mid d_{SA}(v, R_t) \leq \alpha_R\}$ of V, with the selecting probability of a point $v \in N(R_t)$ recalculated as $\mathbb{P}(N_{t+1} = v) = \frac{g_R(v)}{\sum_{u\in N(R_t)} g_R(u)}$. However, it is still unclear how range and rate of switching affect the performance of a SAS strategy, which will be addressed in the next subsection, to better formulate the strategy design problem.

[11] We consider the most powerful culprit (with full knowledge of a strategy) as an extreme case to examine the performance of an SAS system against compromised strategies. In other cases, a weaker culprit can at least observe the long-term visiting probability of any point $v \in V$ as prior knowledge. On the other hand, SAS strategies are required to be disclosed for transparency, such as a crowdsourcing SAS scenario, which increases the risk of a strategy being leaked to culprits.

5.3.2 *Graph Walk: A Chain of Switching Actions*

Over the discrete time span $\mathcal{T}$, any SAS process is now a chain of switching actions in the assignment space V. Recall that a switching sw_t^Y is a relocation (tuning in $\mathcal{S}$ and/or movement in **A**) of a device $Y \in \mathcal{M} \cup \mathcal{R}$ (monitor or culprit[12]) from time $t-1$ to t, whose range is upper-bounded by the switching capacity α_Y. Next, we discuss switching actions from range (how far) and time (how quick) aspects, to formulate the surveillance process into a graph walk.

5.3.2.1 Range Aspect (Switching Capacity)

The switching range refers to the maximum spectra-location distance d_{SA} over which one switching action is possible. As discussed before, switching actions of dedicated monitors are completed via physical movements and/or tuning of individual monitors, so the switching cost scales with spectra-location distance d_{SA}. For this case, we need a quantitative metric to accurately measure the switching cost for strategy design, which is addressed in Sect. 5.4, where a low switching-cost monitoring strategy is proposed for dedicated monitors.

On the other hand, switching in a crowdsourcing monitoring scenario is a change of surrogate monitors, that is, wireless devices (spontaneously) participating in the SAS process. In this case, the switching cost may not scale with the spectra-location distance. Instead, a switching between any two assignment points v_x and $v_y \in V$, can either be possible with a fixed amount of cost (e.g., time and coordination budget), or impossible during one time step. To be more specific, when immediate communication is guaranteed among all participants, a 'handover' between any two monitoring surrogates will be possible in one time step, i.e., $\alpha_M = \infty$; otherwise for distributed crowdsourcing monitoring that relies on short-range wireless communication to coordinate, any switching action is only possible between two monitors within their communication ranges, i.e., $\alpha_M < \infty$. For crowdsourcing monitoring scenarios under unlimited ($\alpha_M = \infty$) and limited ($\alpha_M < \infty$) capacities, strategy design is discussed in Sects. 5.5 and 5.6, respectively.

5.3.2.2 Time Aspect (Switching Rates)

In addition to switching range, the rate of switching, that is, how many switching actions can be conducted in one time step, is also limited by the hardware constraint. Intuitively, a faster-switching culprit will be more difficult to catch, due to the shorter time that the culprit remains in the monitoring power of any monitors.

[12] Normal users in $\mathcal{U}$ can also switch, but their switching actions do not have any impact on the performance of SAS strategies, only the outcome, and are hence not listed here.

Counter-intuitively, as we will show in Lemma 5.1, such culprits will be detected in an even shorter period of time.

In Lemma 5.1, consider spectrum monitors with $q(\delta)$-monitoring power, that is, when a culprit shows up in the cell assigned to a monitor (referred to as *co-location*), the probability that it is detected by that monitor during one time step is q. Let $q \cdot p(s)$ ($s \in [0, 1]$) denote the detecting probability when the co-location time s is less than one full time step, where the non-decreasing function $0 \leq p(s) \leq 1$ captures the attenuated detection probability due to the decreased co-location time, and has the property of $p(0) = 0$, $p(1) = q$.

Lemma 5.1 *Suppose culprit R_1 differs from R_2 in only the switching rate: R_1 can switch $k \in \mathbb{N}^+$ times during one time step, while R_2 and monitors can switch once. The detection time of R_1 is stochastically dominated by that of R_2, that is, for any strategy f,*

$$\tau_{R_1}(f) \overset{d}{\leq} \tau_{R_2}(f), \tag{5.15}$$

when the following criterion is satisfied:

$$p(\frac{1}{k}) \geq \frac{1 - [1 - q \cdot \mathbb{P}\big(R^2(t) \in C(f_t^m)\big)]^{\frac{1}{k}}}{q \cdot \mathbb{P}\big(R^2(t) \in C(f_t^m)\big)}. \tag{5.16}$$

Proof Without loss of generality, suppose the strategy f is carried out by one monitor M. Equation (5.15) holds trivially for a deterministic strategy f_S, when both R_1 and R_2 are adversarial, because every next to visit assignment point is known by an adversarial culprit. In fact, the detection time $\tau_{R_1}(f_S) = \tau_{R_2}(f_S) = \infty$, which is referred to as the 'wandering hole' problem analyzed at the end of Sect. 5.4.

Next we consider the case of f being a randomized strategy, in which the next assignment point is chosen randomly, or R_1 and R_2 being persistent culprits. During a time step t, R_1 generates an exploit sequence $\{R_1^1(t), R_2^1(t), \cdots, R_k^1(t)\}$, and R_2 switches to $R^2(t)$, while monitor M stays at a fixed assignment point $f_t(M) \in V$. The probability that R_2 is detected during t is $Q_2 = q(1 - \mathbb{P}\big(R^2(t) \neq f_t(m)\big))$, while for R_1, the probability of being identified by monitor M is

$$Q_1 = 1 - \Pi_{i=1}^k \left(1 - qp(\frac{1}{k})\mathbb{P}\big(R_i^1(t) = f_t(m)\big)\right). \tag{5.17}$$

Given that R_1 and R_2 only differ in switching rates, that is, $\mathbb{P}\big(R_i^1(t) = v\big) = \mathbb{P}\big(R^2(t) = v\big)$, for any $v \in V$ and $i \in \{1, 2, \cdots, k\}$, the probability $Q_1 = 1 - (1 - p(\frac{1}{k})Q_2)^k$. When the condition in Eq. (5.16) holds, we have $Q_1 \geq Q_2$, which means that R_1 is more probable to be detected during any given time step t. Equivalently, the complementary CDF (CCDF) of detection time $R_1(f)$ and $R_2(f)$ satisfy $\mathbb{P}(\tau_{R_1(f)} > l) \leq \mathbb{P}(\tau_{R_2(f)} > l)$, for any integer $l \geq 1$. □

The condition in Eq. (5.16) easily holds when probability $\mathbb{P}\big(R^2(t) \in C(f_t^m)\big)$ is small, i.e., when the size of the assignment space $|V|$ is large enough, such that it is difficult for a culprit to co-locate with any monitor in a single time step. Otherwise, when the assignment space V contains few assignment points, though catching the faster culprit R_1 takes more time than the slower one, the expected detection time can be derived as $\mathbb{E}(\tau_{R_1}) = \frac{1}{kq \cdot p(\frac{1}{k})\mathbb{P}(R_2 \in C(f_t^m))}$, which will not be large, if the attenuation function satisfies $p(\frac{1}{k}) \geq \frac{1}{k}$. Consequently, the difference between the detection time of R_1 and R_2 will be small. Based on this observation, it is reasonable to assume both culprits and monitors switch once in every time step.

5.3.2.3 A Graph Walk on Composite Graph (G_M, G_R)

Accounting switching capacity, the assignment space V is more than a set of points, rather, a subspace that inherits d_{SA} metric from space X. Together with the space-tessellation procedure, the subspace gives rise to a structure that incorporates the possibility of switching actions, i.e., a composite graph, which consists of:

1. *Monitoring subgraph* $G_M = (V, E_M)$, in which an edge $(u, v) \in E_M$ exists, if and only if $d_{SA}(u, v) \leq \alpha_M$. Then, an arbitrary strategy $\{f_t^m\}_{t \in \mathcal{T}}$ can be seen as a joint *walk* by $m = |\mathcal{M}|$ monitors on the monitoring subgraph G_M.
2. *Exploiting subgraph* $G_R = (V, E_R)$, in which an edge $(u, v) \in E_M$ exists, if and only if $d_{SA}(u, v) \leq \alpha_R$ of the culprit. Then, the exploiting sequence $\{R_t\}_t$ of R, which contains the assignment points exploited by culprit R, also corresponds to visited vertices of a walk on the exploiting subgraph $G_R = (V, E_R)$.

Graph G_R and G_M have the same vertex set V, and are both sub-graphs of K_n, i.e., the complete graph with $n = |V|$ vertices, which corresponds to SAS scenario of unlimited switching capacities ($\alpha_M = \alpha_R = \infty$). The SAS process, particularly culprit detection, then becomes a graph walk on the composite graph $G = (G_M, G_R)$, in which culprit R is first detected when R and any of the monitors in $\mathcal{M}$, co-locate at an assignment point, i.e., meet on a vertex in G.

Figure 5.7 illustrates an example of the composite graph walk. Monitors $\mathcal{M} = \{M_1, M_2\}$ (solid blue dot) and a culprit R (solid red dot) both reside/walk in the assignment space $V = \{a, b, c, d, e, u, v, x\}$ for two time steps, where the edge

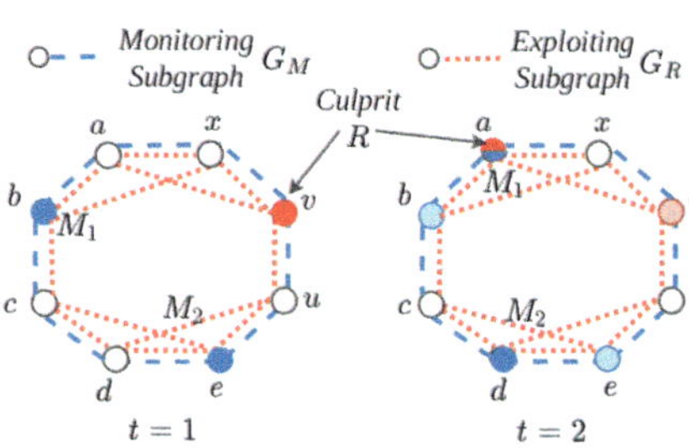

Fig. 5.7 An example of culprit detection process by monitors M_1 and M_2 (blue dots) on the composite graph (G_M, G_R)

Fig. 5.8 Monitoring (green) and exploiting (orange) subgraphs for the simulation scenario in Fig. 5.22, which illustrates the case of weaker (smaller α_M) monitors vs. the more powerful (larger α_R) culprit

sets E_M and E_R are shown in blue dashed and red dotted lines respectively. By $t = 2$, the sweep coverage $C_t = \{a, b, c, e\} \subset V$, indicating the coverage time $T_f > 2$. The detection time $\tau_R(f) = 1$, because monitor M_1 meets with the culprit R_t at assignment point $a \in V$, during time step $t = 2$. Figure 5.8 illustrates the exploiting and monitoring subgraphs in a culprit detection scenario, where the switching capacity of the monitor is weaker than the culprit.

Formulating the SAS process into a graph walk makes strategy design more tractable, in the sense that the strategy space is now discrete and finite, such that both the theoretic and simulation approaches are viable. Under this formulation, we discuss strategy design and performance evaluation for the dedicated and crowd-sourcing monitoring scenarios, in the following Sects. 5.4 and 5.5, respectively.

5.4 Deterministic SAS Strategies for Dedicated Monitors

Dedicated monitors refer to the specialized monitoring equipment mounted on towers, drones, vans, etc. which are widely used by governmental and commercial agents, e.g., FCC, NTIA, and AT&T, to collect spectrum measurement data [13]. For DSA-enabled systems relying on dedicated monitors, *deterministic* monitoring strategies, that is, monitors traverse a predetermined route to sweep-scan the spectra-location space, are the sensible choice due to its simpleness, e.g., [15]. For such strategies, the *switching cost*, in the form of time, energy or budget, is the key concern in deploying dedicated monitors. The reason behind this is that, switching cost, induced by tuning (in spectrum domain **S**) and movement (in space domain **A**), scales with distances in both domains, so it is essential to optimize the strategy for a reduced cost, given that all the monitors are under the control of the SAS function. Therefore, we first define a comprehensive switching cost metric, based on the optimization of which, we propose the low-cost deterministic SAS strategy f_S.

Definition 5.7 The **switching cost** from point x_i to $x_j \in X$, is defined as the sum of tuning cost and relocation cost, that is,

$$\gamma(x_i, x_j) := \beta_S d_S(p_S(x_i), p_S(x_j)) + \beta_A d_A(p_A(x_i), p_A(x_j)), \tag{5.18}$$

where β_S and β_A are cost coefficients for tuning and relocation respectively. The **cost of strategy** $\{f_t^m\}_{t=1}^{T_f}$ is then

$$\Gamma_f^m := \sum_{t=2}^{T_f^m} \sum_{i=1}^{m} \gamma(f_t^m(M_i), f_{t-1}^m(M_i)), \tag{5.19}$$

where T_f^m is the coverage time defined in Eq. (5.5).

Switching cost Γ_f can be applied to describe time, energy, and budget. For example, the cost coefficients can be set to $\epsilon\beta_A >> \beta_S$ to describe the switching time, since the time it takes a radio head to tune to a different center frequency is approximately 1 ms [16], during which the physical movement of any mobile device is negligible. It could also be the case that re-configuring the center frequency of a radio head is more expensive (in terms of budget) than physical movements, e.g., specialized devices with narrow frequency ranges, such that $\epsilon\beta_A < \beta_S$. Accounting for switching cost between assignment points in space V, the resulting monitoring sub-graph G_M becomes a weighted complete graph K_n, in which the weight of each edge reflects the switching cost along that edge.

5.4.1 *Low Cost Deterministic Strategies f_S*

Addressing the design concern of switching cost in the dedicated SAS scenario, an *optimal* deterministic strategy is a strategy $\{f_{S,t}\}_{t=1}^{\mathcal{T}}$, whose total switching cost Γ_f is minimized in the solution space $\{V^m\}^{\mathcal{T}}$. In this sense, finding an optimal strategy $\{f_{S,t}^m\}_{t=1}^{T_s}$ is equivalent to finding m vertex-disjoint 'shortest' paths (in terms of switching cost) that jointly cover the assignment space V by time $T_S^m = \lceil \frac{n}{m} \rceil$. This problem is actually an open-path multi-depot multi-travailing salesmen problem (MD-MTSP), which is known to be NP-hard [5, 28].

Observing the structure of assignment space V (as shown in Fig. 5.5) over a rectangular region $\mathbf{A}$, we prove there is an upper bound of the minimum switching cost Γ_{min}^m for small m values, which can be achieved by traversing the space V 'smartly'. Before presenting the main result, we summarize necessary denotations in the following Table 5.3 for the ease of references.

Theorem 5.1 *For a set of $m \geq 3$ monitors on the assignment space V with parameters L, D and H (defined in Table 5.3), the switching cost Γ_{min}^m can be upper bounded by*

$$\Gamma_{min}^m \leq \begin{cases} \Gamma_*^1 - A_*(m-1), & \text{if } m = 2k+1, \\ \Gamma_*^2 - A_*(m-2), & \text{if } m = 2k+2, \end{cases} \tag{5.20}$$

where $k \in \mathbb{N}^+$, $* = f$ if $\beta_S \geq \frac{2-\sqrt{2}}{2}\beta_A\epsilon$, and $* = g$ otherwise. Quantities Γ_*^1, Γ_*^2 and $A_*(\cdot)$ are determined as

$$\Gamma_*^1 := K_1\gamma_1 + K_2^*\gamma_2 + K_3^*\gamma_3, \tag{5.21}$$

$$\Gamma_f^2 := \Gamma_f^1 - (2D-1)\gamma_2 + \gamma_3 + (\frac{H}{2}\gamma_1 + \gamma_2)e(L), \tag{5.22}$$

$$\Gamma_g^2 := \Gamma_g^1 - 2\gamma_2 - \gamma_1 + 2\gamma_3, \tag{5.23}$$

$$A_*(x) := \min\{K_1, x\}\gamma_1 + \min\{K_3^*, [x-K_1]^+\}\gamma_3 + [x - K_1 - K_3^*]^+\gamma_2, \tag{5.24}$$

where $K_1 = LD(H-1)+(L-1)(D-1)(H-2)$, $K_2^f = (L-1)D+(L-2)(D-1)$, $K_3^f = 2(D-1)$, $K_2^g = L+D-2$, $K_3^g = 2(L-1)(D-1)$, and $e(L) = 1$ when L is even, $e(L) = 0$ otherwise.

Proof *(Sketch)* We prove this theorem in a constructive manner, and work our way from the base case when ($m = 1$ or 2) toward the general cases ($m \geq 3$). For the single- or double-monitor cases ($m = 1$ or 2), considering that Type-1, 2 and 3 edges are the least expensive edges in the complete graph K_n in terms of switching cost, the main idea is to construct a traversing route with the most number of least expensive edges (Type-1 or Type-2 depending on quantity $\beta_S - \frac{2-\sqrt{2}}{2}\beta_A\epsilon$), and as few necessary longer edges as possible. The optimal strategy for these two cases, and their accompanied minimum switching costs Γ_{min}^1 and Γ_{min}^2, are presented in Lemmas 5.2 and 5.4, respectively. Then for $m > 2$, the switching cost can be upper bounded by evenly dividing the low-cost routes of $m = 1$ or $m = 2$, and then removing the induced dividing edges, as presented in Lemma 5.3. □

Lemma 5.2 *When $m = 1$, and the assignment space V has L, D and H assignment points along the length, width and height (spectrum $\mathcal{S}$) of X, the switching cost of the optimal strategy Γ_{min}^1 satisfies*

$$\Gamma_{min}^1 \leq \Gamma_*^1 = K_1\gamma_1 + K_2\gamma_2 + K_3\gamma_3, \tag{5.25}$$

Table 5.3 Parameters and denotations in Theorem 5.1

Denotation	Meaning	Example in Fig. 5.5
L	# of assignment points along the length of $\mathcal{A}$	$L = 3$ (in Fig. 5.5 right)
D	# of assignment points along the width of $\mathcal{A}$	$D = 3$ (in Fig. 5.5 right)
H	# of assignment points along the spectra axis $\mathcal{S}$	$H = 4$ (in Fig. 5.5 right)
$n = \|V\|$	Total # of assignment points in V	$n = 48$ (in Fig. 5.5 right)
Type-1 edge	Edges w/ cost $\gamma_1 := \beta_S a$	(c_1, c_6), (c_2, c_7) (in Fig. 5.5 left)
Type-2 edge	Edges w/ cost $\gamma_2 := \beta_A\epsilon a$	(c_1, c_2), (c_7, c_8) (in Fig. 5.5 left)
Type-3 edge	Edges w/ cost $\gamma_3 := \frac{a}{2}(\sqrt{2}\beta_A\epsilon + \beta_S)$	(c_1, c_5), (c_5, c_8) (in Fig. 5.5 left)

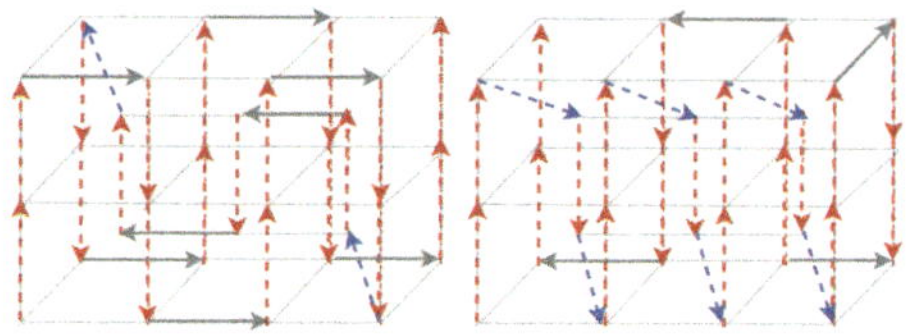

Fig. 5.9 Low-cost strategies for $m = 1$ monitor ($f^*_{m=1}$ and $g^*_{m=1}$). Red, grey and blue indicate Type-1, 2, and 3 edges (defined in Table 5.3), respectively

where $K_1 = LD(H-1) + (L-1)(D-1)(H-2)$*;* K_2 *and* K_3 *are determined as:*

1. *When* $\beta_S \geq \frac{2-\sqrt{2}}{2}\beta_A\epsilon$, $K_2 = (L-1)D + (L-2)(D-1)$ *and* $K_3 = 2(D-1)$*. Then* Γ^1_* *is achieved by the strategy* $f^*_{m=1}$*, as shown in Fig. 5.9 (left).*
2. *Otherwise* $K_2 = L + D - 2$ *and* $K_3 = 2(L-1)(D-1)$*, achieved by strategy* $g^*_{m=1}$*, as shown in Fig. 5.9 (right).*

Proof The shortest (in terms of cost) combination of paths connecting the $n = |V|$ points is including all the vertical (Type-1) edges, and then adding the minimum number of necessary type-3 and type-2 edges, as in Fig. 5.9, because the triangle equality holds under the switching cost metric defined in Sect. 5.4. When $\beta_S \geq \frac{2-\sqrt{2}}{2}\beta_A\epsilon$, the switching cost of the optimal strategy $\Gamma^1_{min} = \Gamma^1_{f^*}$ since the short edges (Type-1 and Type-2) are utilized to the most, indicating $f^*_{m=1}$ (Fig. 5.9 left) is the optimal strategy because in this case $\gamma_2 \leq \gamma_3$. Otherwise, if $\gamma_2 > \gamma_3$, it is cheaper to use Type-3 edges than Type-2 edges more often, under the circumstance that the total number of edges remains the same, i.e. $K_1 + K_2 + K_3 = n - 1$. Therefore strategy $g^*_{m=1}$ (Fig. 5.9 (right)) achieves a lower switching cost than strategy $f^*_{m=1}$. □

Now we consider the case of multiple monitors, i.e., $m > 1$.

Lemma 5.3 *When* $m > 1$*, the minimum cost* Γ^m_{min} *is upper-bounded by*

$$\Gamma^1_{f^*} - \min\{K_1, (m-1)\}\gamma_1 - \min\{K_3, [m-1-K_1]^+\}\gamma_3 - [m-1-K_1-K_3]^+\gamma_2, \tag{5.26}$$

where $[x]^+ = \max\{x, 0\}$*, and* K_1*,* K_3*,* Γ^1_{min} *defined as in Lemma 5.2.*

Proof One possible solution of employing $m > 1$ monitors is to divide the near-optimal path $f^*_{m=1}$ in Lemma 5.2 evenly (in terms of hop count), into m sub-paths, by removing $m-1$ edges. Note that different m will result in the removal of different numbers of Type-1, 2 and 3 edges. However, the minimum cost Γ^m_{min} is always upper bounded by the case that the removal starts from Type-1 edges, which are associated with the minimum switching cost. □

Further, when m is even, a tighter bound can be obtained through an effective way to divide the optimal path in strategy $f^*_{m=1}$ and $g^*_{m=1}$, as shown in Lemma 5.4.

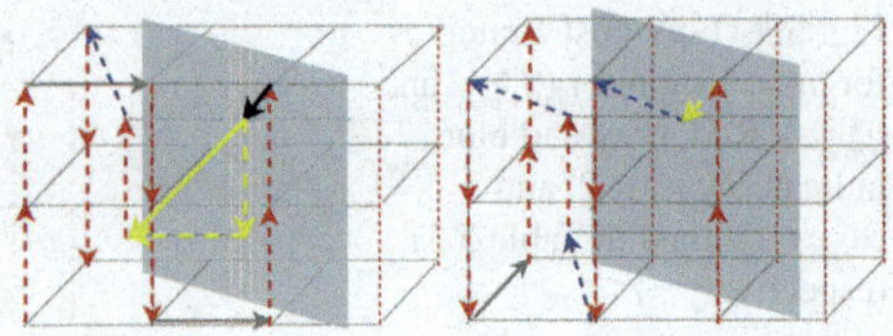

Fig. 5.10 Low-cost strategies for $m = 2$ monitor ($f^*_{m=2}(M_1)$ and $g^*_{m=2}(M_1)$)

Lemma 5.4 *When $m = 2$, then*

$$\Gamma^2_{g^*} = \Gamma^2_{g^*} - 2\gamma_2 - \gamma_1 + 2\gamma_3, \tag{5.27}$$

$$\Gamma^2_{f^*} = \Gamma^2_{f^*} - (2D - 1)\gamma_2 + \gamma_3 + (\frac{H}{2}\gamma_1 + \gamma_2)e(L), \tag{5.28}$$

where $e(L) = 1$ if L is even, and $e(L) = 0$ otherwise. The minimum switching cost satisfies $\Gamma^2_{min} \leq \min\left\{\Gamma^2_{f^*}, \Gamma^2_{g^*}\right\}$. Further if the number of monitors $m = 2k$ and $k \geq 2$, then $\Gamma^m_{min} \leq \Gamma^2_{min} - (m-2)\gamma_1$.

Proof Equation (5.27) is achieved by evenly dividing the space into two parts, and augment the divided path segment of $f^*_{m=1}$ and $g^*_{m=1}$ by adding extra Type-3 edges, as shown in Fig. 5.10. For example, in $f^*_{m=2}$, a Type-3 edge (black arrow in Fig. 5.10 left) is added for each row. When L is even, the middle line of Type-1 edges needs to be divided, so a longer edge (solid yellow arrow) is added. Since the cost metric γ is a linear combination of d_S and d_A, the cost of the solid yellow arrow is equal to the summation of costs $(\frac{H}{2}\gamma_1 + \gamma_2)$ of the two dashed yellow arrows. Similarly for $g^*_{m=2}$ (Fig. 5.10 (right)), two Type-2 edges and one Type-1 edges are removed, while one Type-3 edges are added. Then the minimum switching cost Γ^2_{min} is upper bounded by the minimum of the two. For $m = 2k \geq 4$, applying Lemma 5.3 to Γ^2_{min} yields the result. □

Combining Lemmas 5.2, 5.3 and 5.4, we have the upper-bounds of the total switching cost Γ_f in Theorem 5.1.

As a constructive proof, Theorem 5.1 also sketches the proposed deterministic strategy (denoted as f_S). To validate these results (strategy f_S and upper bounds), we conduct numerical simulation, whose configuration is enumerated in Table 5.4. Note that the width of the spectrum block S does not have any unit (similarly for monitoring power parameter δ). We eliminate the unit rather than plugging in parameters of real-world hardware e.g., [23, 25], because there will be a (mere) change of constant when different units are applied, which is insignificant (and more confusing) in validating the efficiency of the proposed deterministic strategies.

The proposed strategy (black bars on the far left in Figs. 5.11 and 5.12, labeled as f_S) is compared with the best solutions found by genetic algorithm (red-toned bars in Fig. 5.11, labeled as GA-x, which means 10^x iterations are executed to obtain this result), which is a commonly used heuristic for MTSP, and those found by a greedy-based random search (blue-toned bars in Fig. 5.12, labeled as RS-x, which means

Table 5.4 Simulation configuration for coverage time of strategy f_S^m

Parameter	Description	Value
$n = \|\mathcal{V}\|$	Number of assignment points	394
$m = \|\mathcal{M}\|$	Number of monitors	[1, 5]
$\mathcal{A}$	The geographical region of interest	$[0, 10]^2$
$\|S\|$	Width of spectrum block	18
ϵ	Scaling coefficient in distance d_{SA}	5
δ	Radius of the monitoring power	$\frac{\sqrt{5}}{2}$
(L, D, H)	Tessellation parameters (see Table 5.3)	(5,5,10)
β_S, β_A	Switching cost coefficients (see Definition 5.7)	(0, 1)

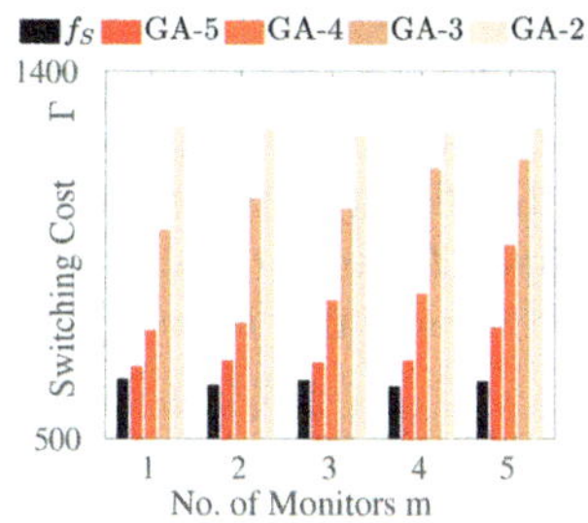

Fig. 5.11 The proposed deterministic strategy f_S (black on the far left) achieves a lower switching cost, compared to the genetic algorithm (GA) solution (red bars) under $\beta_S = 0.8$, $\beta_A = 0.2$

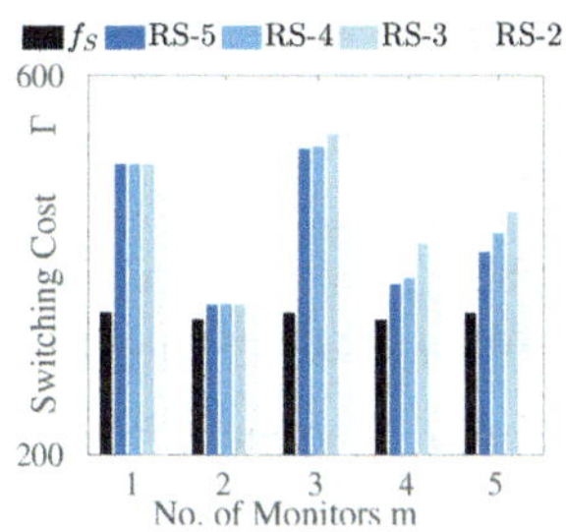

Fig. 5.12 The proposed deterministic strategy f_S (black on the far left) achieves a lower switching cost, compared to the greedy-based random search (RS) solution (blue bars) under $\beta_S = 0.1$, $\beta_A = 0.9$

10^x iterations are executed to obtain this result). The proposed f_S achieves a very close switching cost to that of the best solution/strategy output by GA after more than 10,000 iterations, under different switching cost coefficient settings. In fact, when m is small, the best solution provided by GA bears great resemblance to the proposed strategy, in terms of traversed edge types, as well as the breaking points (i.e., the way to divide the optimal path of $m = 1$). Similar observations can be obtained in the comparisons with random search, despite that the optimal solution can not be easily found by the random search due to its greedy nature.

5.4.2 *Detection Time of the Deterministic Strategy f_S*

Though deterministic strategy f_S proposed in Theorem 5.1 proves to be a good strategy in terms of coverage time ($T_S^m = \lceil \frac{|V|}{m} \rceil$) and switching cost, its detection performance is not satisfying. In fact, it may suffer from a 'wandering hole' problem, when adversarial spectrum culprits are present. So in this subsection, we discuss the detection performance of the deterministic strategy f_S.

5.4.2.1 Detecting Persistent Culprits R_p

The expected detection time $\mathbb{E}(\tau_p(f_S))$ of different persistent culprits R_p under strategy f_S can be found in Table 5.2, given that the initial position of R_p (at time $t = 0$) is chosen randomly from the assignment space V. As can be seen, the expected detection time $\mathbb{E}(\tau_p(f_S))$ are bounded above by the coverage time $T_S^m = \lceil \frac{n}{m} \rceil$. Proofs of these results are listed as follows.

Proof *(Expected Detection Time Presented in Table 5.2)* In general, for a strategy $\{f_t^m\}_{t\in\mathcal{T}}$ conducted by a set $\mathcal{M}$ of monitors with $q(\delta)$-monitoring power, the probability that a culprit R_p is detected in time step $t = k$ is $p_k := q \sum_{v\in f_k(\mathcal{M})} g_{R_p}(v)$. Since R_p is persistent, i.e., its exploiting pattern does not change over time, the probability that it not detected until time step k can be derived as $\mathbb{P}\big(\tau_{R_p}(f_t) = k\big) = p_k \prod_{i=0}^{k-1}(1 - p_i)$, then the expected detection time of a persistent culprit R_p with PMF $g_{R_p}(v)$ can be obtained by

$$\mathbb{E}\big(\tau_p(f)\big) = \sum_{k=1}^{\infty} k p_k \prod_{i=1}^{k-1}(1 - p_i), \tag{5.29}$$

where $p_0 = 0$ is added to keep its form consistent. Next, we discuss the detection time of the three types of persistent spectrum culprits.

Case 1 Stationary R_s without DSA capability, who chooses $x \in X$ (or equivalently $v \in V$) uniformly at random, and stays there for any $t > 0$. Therefore, the probability that this culprit is detect at time step k is $\mathbb{P}(\tau_{R_s}(f_S) = k) = \frac{1}{T_S}$, indicating a uniform distribution for $k = 1, 2, \ldots, s$, and $\mathbb{P}(\tau_R(f_S) > T_S) = 0$. The expected detection time $\mathbb{E}(\tau_{R_s}(f_S)) = \frac{T_S}{2}$ is bounded by the coverage time of deterministic strategy f_S.

Case 2 Stationary R_{sd} with DSA capability, that fixes its location $a \in \mathcal{A}$, and hops to any frequency chosen uniformly at random from $\mathcal{S}$, resulting in a $R_{sd}(t)$ such that $p_a(R_{sd}(t)) = a$ for all $t > 0$. Exact analysis of $\mathbb{E}(\tau_{R_{sd}}(f_S))$ is fairly difficult, due to the slight overlap of monitoring powers at location a. But for every time slot $t = k$, the probability that culprit R_{sd} is identified, is strictly less than that the more capable R_{md} is identified, i.e., $\mathbb{P}(\tau_{R_{sd}}(f_S) = k) < \mathbb{P}(\tau_{R_{md}}(f_S) = k)$. Therefore, $\mathbb{E}\left(\tau_{R_{sd}}(f_S)\right) \leq \mathbb{E}\left(\tau_{R_{md}}(f_S)\right)$.

Case 3 Mobile R_{md} with DSA capability, whose exploit sequence is randomly chosen from X (or equivalently discrete space V) during every time step. Since there are m monitors in the assignment space V, during every time step t, the probability that the culprit R_{md} is not detected is $1 - p = \frac{n-m}{n}$. Consequently, r.v. $\tau_R(f_S)$ is geometrically distributed with parameter p, that is, $\mathbb{P}(\tau_R(f_S) = k) = (1-p)^{k-1}p$. Hence its expected value $\mathbb{E}(\tau_R(f_S)) = \frac{n}{m} \simeq T_S$.

□

5.4.2.2 Detecting an Adversary Culprit R_a

To maintain a low switching cost, the deterministic strategy f_S will be repeated (in forward/reverse direction in odd/even cycles) after $T_S = \lceil \frac{n}{m} \rceil$ time. Consequently, it is possible for culprits with learning capabilities, i.e., adversarial culprits, to observe the deployment pattern of monitors, and even *predict* where monitors will not be (e.g., the coverage hole shown in Fig. 5.3 right) in the next time step. Then culprits can continue chasing the 'hole' for an infinitely long time, as if hiding in a 'wandering hole' of the dynamically changing coverage $C_t(f_{S,t})$.

Definition 5.8 Strategy $\{f_t\}_{t\in[1,T]}$ is said to suffer from a **wandering hole** problem, if an adversarial culprit can exploit the system for infinitely long time, that is, the expected detection time $\mathbb{E}\big(\tau_a(f_t)\big) = \infty$.

An example of the wandering hole problem is shown in Figs. 5.13 and 5.14, where activities in spectrum block $\mathcal{S}$ over region $\mathcal{A} = [0, 4]^2$ are being monitored by $m = 5$ monitors. Red dots indicate the assignment points of V, calculated by the Kelvin structure tessellation (in Sect. 5.3), while space enclosed by shaded spheres

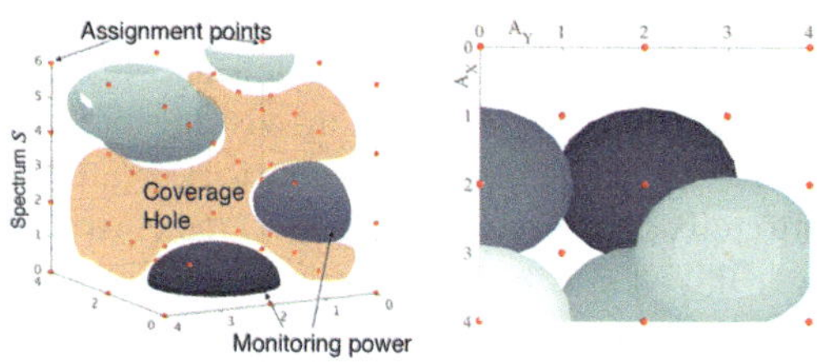

Fig. 5.13 Five monitors are deployed in region $\mathcal{A} = [0, 4]^2$ with $\delta = \frac{\sqrt{5}}{2}$. Their coverage C_t at time $t = 1$ (side-view and top-view), is illustrated as the enclosed space of the blue (partial) balls and the boundary

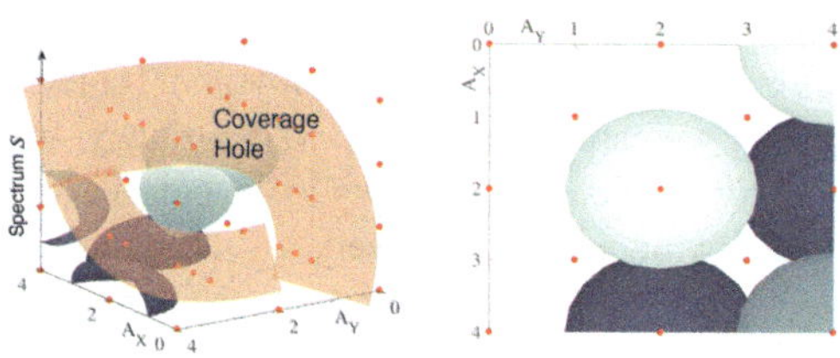

Fig. 5.14 At time $t = 2$, the coverage C_2 changes due to the switching actions of the monitors. The unmonitored space ('hole') is changes in X over time, creating a *wandering hole* across the spectral-location space

corresponds to the monitoring power of the deployed monitors,[13] whose tuning frequency is identified by the darkness of the shade. The white space outside of these spheres corresponds to spectrum slices and locations that a culprit can exploit without being detected, i.e. a spectrum *hole* in the monitoring coverage C_t. At time $t = 1$, consider an adversarial culprit R_a located at (3, 3) occupying lower frequency portion in the 'hole'. From previous observation, culprit R_a can easily find spectra-location points to exploit in the next time slot $t = 2$, because a deterministic monitoring strategy f_S repeats itself periodically, i.e., Void(2) identifies the 'hole' exactly as the 3-D region in X where the probability for it to be monitored during $t = 2$ equals to zero. Consequently, R_a can safely stay at the current location, and continue occupying the current spectrum slice without being detected. In other words, adversarial R_a can swiftly hide in the 'wandering hole' indefinitely, unless the deterministic deployment strategy f_S changes.

Remark 5.2 In fact, the wandering hole problem exists in any deterministic strategy. Once the SAS strategy (or more precisely, its probability distribution) is known by an adversarial culprit R_a, this prior knowledge can be leveraged by R_a to actively dodge monitors, whenever there is a hole in the current coverage, i.e., $X \setminus C_t \neq \phi$.

5.5 Patching the 'Wandering Hole': Randomized Strategies

The wandering hole problem exposes a defect of deterministic strategies against adversarial culprits R_a. The root cause of this defect is that, taking advantage of its prior knowledge, i.e., the *difference* among visiting probabilities on different assignment points, an adversarial culprit R_a is able to determine the spectrum 'hole' to exploit, i.e., Void(t). The sharper the difference, the clearer the boundary of the 'hole', and the larger the chance for the culprit R_a to dodge monitors in the next time step. For instance, under the deterministic strategy f_S shown in Figs. 5.13 and 5.14, a culprit R_a located at $a = (3, 2)$ can easily identify Void(2) $\subset \mathcal{S} \times \{a\}$ due to the prominent difference in probability (the second 'bar' from the left in Fig. 5.15).

Knowing the root cause of the 'wandering hole' problem, a straight-forward countermeasure is to better protect, or frequently change the SAS strategy, such that obtaining visiting probabilities, or the strategy implementation, are more difficult to culprits, which requires constant effort in the deployment stage. Nonetheless, we can achieve the same goal if we carefully design a SAS strategy, from which no useful 'knowledge' can be derived, even if it is known to the culprit. In other words, we can fully *randomize* the deployment, such that every assignment point is visited with the same probability in the long run. Consequently, there is no probability difference,

[13] Since we are interested in the closed space $\mathcal{S} \times \mathcal{A}$, spheres indicating the boundary of monitoring power are trimmed, when they intersect with the boundary of space $X = \mathcal{S} \times \mathcal{A}$.

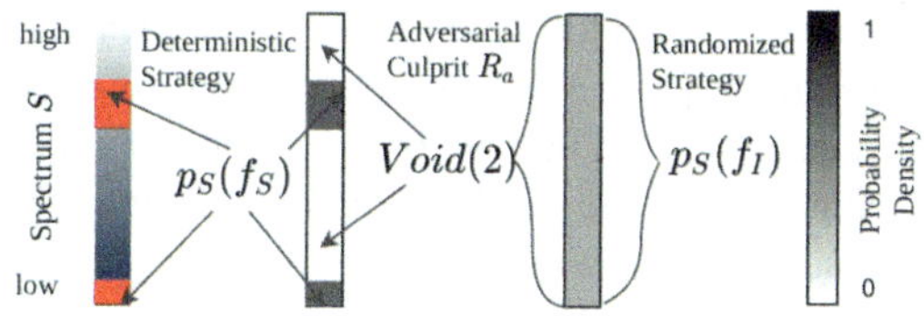

Fig. 5.15 Root cause of the *wandering hole* problem: difference in visiting probability (density)

and hence no boundary of spectrum holes for culprits to locate, as illustrated by the (uniform) grey bar, second from the right in Fig. 5.15. In addition, monitors follow no pattern at all, when switching to a different assignment point, so the adversarial culprits essentially become persistent ones, as their prior knowledge is nullified. Therefore, these randomized strategies will be effective in culprit detection.

On the other hand, as we shift focus from low-cost sweep-coverage to quick culprit detection, we switch gear from switching cost (Definition 5.7) to switching capacity (Definition 5.2) - the latter can be viewed as a binary quantification of the former against a cost threshold. As a result, the underlying graph (G_M, G_R) (defined in Sect. 5.3.2.3) changes from complete graphs with numerical edge weights (cost), to sparser simple graphs without edge weights, pruned according to the switching capacity of monitors and culprits. This change is also consistent with the SAS scenario change (from dedicated to crowdsourcing[14]): considering a switching action is actually a change of surrogate monitors, randomized strategies are natural for the crowdsourcing monitoring scenario, where the switching cost does not scale with distance d_{SA}. To be more specific, with a centralized coordinator or guaranteed communication among participants, switching between surrogate monitors can be timely coordinated, so the entire assignment space V is contained in the switching capacity of all monitors, which translates to the unlimited switching capacity case, i.e., $\alpha_M = \infty$; in case of distributed control, when change of surrogate monitors needs to be completed with local communication, the switching capacity of monitors will be upper-bounded, i.e., $\alpha_M < \infty$, due to their limited communication ranges.

Next, we introduce two simple, but effective randomized strategies.

5.5.1 *Randomized SAS Strategies*

We discuss two strategies that require different levels of coordination and switching capacities: the independent I-strategy f_I and the distributed D-strategy f_D.

1. *I-strategy* f_I. During each time step, each monitor $M_i \in \mathcal{M}$ switches to point $v_i \in V$ uniformly at random, and independently of others, within its switching capacity α_M. The surveillance process is then equivalent to a

[14] For SAS with dedicated monitors, this modeling approach with switching capacity is also valid, if we determines switching capacity $\alpha_M < \infty$ according to a fixed switching cost threshold γ^*.

composite random walk of $m = |\mathcal{M}|$ walkers, each independently generating a sequence $\{f^m_{t,I}(M)\}_{t\in\mathcal{T}}$, on the monitoring subgraph G_M. When G_M is (close to) regular, i.e., the number of assignment points reachable in one switching action is almost the same starting from every point in V, the uniform transition probability leads to a convergence-guaranteed distribution of visiting probabilities, that is, $\pi_v = \frac{1}{n}$, $\forall v \in V$.

2. *D-strategy* f_D. Assignment space V is evenly divided into m disjoint subsets $\{V_i\}_{M_i\in\mathcal{M}}$, such that points in each subset V_i are within monitors' switching capacity α_M, and sets in collection $\{V_i\}_{M_i\in\mathcal{M}}$ compose a partition of V. During time step t, each monitor $M_i \in \mathcal{M}$ switches to point $f^m_{t,D}(M_i)$, chosen uniformly at random from its own subset V_i, which contains $n_m = \lceil \frac{n}{m} \rceil$ assignment points. Thus, the D-strategy is equivalent to m independent single-walker random walks, each on a smaller complete graph K_{n_m}. Moreover, graph K_{n_m} is regular, so the stationary visiting probabilities are also uniform.

Based on this design, we first discuss the basic case of $\alpha_M = \alpha_R = \infty$, i.e., SAS strategy without switching constraint, and leave the more complicated $\alpha_M < \infty$ case for the next section. As we will show, both the coverage time and detection time of the two randomized strategies: (1) are bounded, indicating their efficacy; and (2) scale as $O(\frac{1}{m})$ with respect to the number of monitors m, revealing their efficiency.

5.5.2 Coverage Time of the Two Randomized Strategies f_I and f_D

Because of the unlimited switching capacity of both monitors and culprits, the underlying monitoring subgraph G_M are complete graphs, and both the I- and D-strategy are equivalent to random walks, on K_n and K_{n_m}, respectively. The coverage time T^m_I and T^m_D become well-defined r.v.'s that take value in $[1, \infty)$, and their expected values $\mathbb{E}(T_*)$ are referred to as the *cover time* [3]. This quantity is well-studied for a single walker case, which corresponds to the single monitor case (I-strategy and D-strategy are exactly the same, i.e., $f^1_I = f^1_D$), but considering that it is helpful for the latter cases, we present it in Lemma 5.5 for completeness reasons.

Lemma 5.5 *The expected coverage time of strategy f^1_I (or equivalently f^1_D) on the assignment space V is $\mathbb{E}(T^1_I) = n\mathcal{H}_n$, where $\mathcal{H}_n = \sum_{i=1}^n \frac{1}{i}$ denotes the n-th Harmonic number, and $n = |V|$ is the total number of assignment points.*

Proof The surveillance sequence generated by strategy f^1_I (f^1_D) is a single-walker random walk on K_n. Let r.v. $T_i \geq 0$ denote the time interval between the first hitting time of the $i-1$-th vertex and that of the i-th vertex, for any $i \geq 2$. Then T_i is geometric distributed with $p_i = \frac{n-i+1}{n}$, and $T_1 = 1$. So the expected coverage time can be calculated as $\mathbb{E}(T^1_I) = \mathbb{E}\left(\sum_{i=1}^n T_i\right) = \sum_{i=1}^n \frac{n}{n-i+1} = n\mathcal{H}_n$. □

Next, we discuss the case of m monitors for the I-strategy and D-strategy.

5.5.2.1 Coverage Time of the I-Strategy T_I^m

For an I-strategy f_I^m carried out by m monitors, the expected coverage time $\mathbb{E}(T_I^m)$ can be bounded by the following theorem.

Theorem 5.2 *For a set of $m = |\mathcal{M}|$ monitors that follow the I-strategy $\{f_t^m\}_{t\in\mathcal{T}}$ in the assignment space V, the expected coverage time is upper bounded by*

$$\mathbb{E}(T_I^m) \le e(n-1)\left[0.562 + 0.768\frac{\mathcal{H}_n}{m}\right], \tag{5.30}$$

where $n = |V|$ is the number of assignment points in V.

Proof When $m = 1$, the expected coverage time $\mathbb{E}(T_I^1) = n\mathcal{H}_n$ from Lemma 5.5. For $m > 1$, let $T_{I,i}^1$ denote the coverage time of a single monitor $M_i \in \mathcal{M}$. Then $\{T_{I,i}^1\}_{i=1}^m$ is a set of i.i.d. r. v.'s with $T_{I,i}^1 \stackrel{d}{=} T_I^1$ for any monitor $M_i \in \mathcal{M}$. So

$$\mathbb{E}(T_I^m) \le \mathbb{E}(\min_{1\le i\le m} T_{I,i}^1) \le \mathbb{E}(T_{I,i}^1) = \mathbb{E}(T_I^1). \tag{5.31}$$

Let $H(x, y) := \min_{t>0}\{f_t^m(M_i) = y \mid f_0^m(M_i) = x\}$ denote the first hitting time of monitor $M_i \in \mathcal{M}$ on assignment point $y \in V$, given that M_i started its walk (surveillance) from $x \in V$. Since monitor M_i is inter-changeable with M_j, the notation of monitor (i) can be suppressed in the hitting time. Each monitor walks independently on the complete graph K_n, so $\mathbb{E}(H(x, y)) = n - 1$, for any $x, y \in V$.

Suppose at time $t = 0$ (one step ahead of the SAS process starts), all monitors are assigned to the same point x, that is, $f_0^m(M_i) = x$ for all $M_i \in \mathcal{M}$. Then for a fixed point $u \in V$, the probability that a random walk of length $e(n-1)$ does not hit u is upper bounded by the Markov Inequality [3]:

$$\mathbb{P}\big(H(x, u) > e(n-1)\big) \stackrel{\text{Markov}}{\le} \frac{E(H(x, u))}{e(n-1)} = e^{-1}. \tag{5.32}$$

Then for any integer $r > 1$, probability $\mathbb{P}\big(H(x, u) > er(n-1)\big) \le e^{-r}$, since the entire walk can be viewed as r individual trails trying to hit u simultaneously. Therefore, the probability that m independent (single-walker) random walks all starting from x, have not hit point u up to time $t = er(n-1)$, can be upper-bounded by

$$\mathbb{P}\Big(\min_{M_i\in\mathcal{M}}\{H_i(x, u)\} > er(n-1)\Big) \le e^{-mr}. \tag{5.33}$$

Let $r = \lceil \frac{\ln n + \gamma}{m} \rceil$, where $\gamma = \lim_{n\to\infty}(\mathcal{H}_n - \ln n)$ is the Euler-Mascheroni constant. Then the probability that the m random walkers have not covered all points in V by time $er(n-1)$, or equivalently the coverage time T_I^m is greater than $er(n-1)$ can also be upper bounded:

$$\mathbb{P}\big(T_I^m > er(n-1)\big) \leq e^{-mr} \leq e^{-\gamma}. \tag{5.34}$$

Also, notice that the expected values $\mathbb{E}(T_I^m) \leq \mathbb{E}(T_I^1) = n\mathcal{H}_n$ from Eq. (5.31). Therefore, the expected coverage time can be obtained as

$$\begin{aligned}\mathbb{E}(T_I^m) &\leq er(n-1)\cdot(1-e^{-\gamma}) + \mathbb{E}(T_I^1)\cdot e^{-\gamma} \\ &\leq \frac{n-1}{m}\left[(\ln n + \gamma + m)(e - e^{1-\gamma}) + \mathcal{H}_n e^{-\gamma}\right] \\ &\leq \frac{e(n-1)}{m}\left[(\mathcal{H}_n + m)(1 - e^{-\gamma}) + \mathcal{H}_n e^{-(1+\gamma)}\right],\end{aligned} \tag{5.35}$$

and plugging in values of γ and e yields the result. □

Though not a tight bound, Theorem 5.2 reveals the scaling law of the expected coverage time with respect to size n of space V, and number of monitors m, that is, $\mathbb{E}(T_I^m) = O(\frac{n \ln n}{m})$.

5.5.2.2 Coverage Time of the D-Strategy T_D^m

The expected coverage time of D-strategy f_D^m can also be bounded above with the help of Lemma 5.5, as presented in the following theorem.

Theorem 5.3 *For a set of m monitors following the D-strategy f_D^m on the assignment space V of size n, the expected coverage time $\mathbb{E}(T_D^m)$ is upper bounded by*

$$\mathbb{E}(T_D^m) \leq n_m \mathcal{H}_{n_m} + \frac{n_m\sqrt{m-1}}{2(n_m-1)}\left[7(n_m)^2 - 11n_m + 2\right]^{\frac{1}{2}}, \tag{5.36}$$

where $n_m = \lceil \frac{n}{m} \rceil$.

Proof Let r.v. T_{M_i} denote the time when monitor M_i has covered every assignment point in its own subset $V_i \subset V$ for the first time, where $|V_i| = \lceil \frac{n}{m} \rceil =: n_m$, and we refer to T_{M_i} as the i-th sub-coverage time. Under the D-strategy f_D^m, the sequence of assignment points to be scanned by monitor M_i, can be viewed as a trail generated by a random walker on the complete graph K_{n_m} of size n_m. Therefore, by Lemma 5.5, the expected sub-coverage time $\mathbb{E}(T_{M_i})$ by monitor $M_i \in \mathcal{M}$ is $\mathbb{E}(T_{M_i}) = n_m$.

In addition to the mean, we know the distribution of every T_{M_i}, particularly its mean and variance, from the proof of Lemma 5.5, because r.v. $T_{M_i} = \sum_{k=1}^{n_m-1} T_k$, where each T_k is geometrically distributed with parameter $\frac{n_m-k+1}{n_m}$. Hence $\mathbb{E}(T_k) = \frac{n_m}{n_m-k+1}$ and $Var(T_k) = \frac{n_m(k-1)}{(n_m-k+1)^2}$. Then the variance of r.v. T_{M_i} can be derived as

$$
\begin{aligned}
Var(T_{M_i}) &= \sum_{k=1}^{n_m} Var(T_k) = \sum_{k=1}^{n_m} \frac{n_m(k-1)}{(n_m - k + 1)^2} \\
&\leq (n_m)^2 \left(1 + \sum_{j=2}^{n_m} \frac{1}{j^2 - 1}\right) = (n_m)^2 \left(\frac{7}{4} - \frac{2n_m + 1}{2(n_m - 1)n_m}\right).
\end{aligned}
\tag{5.37}
$$

The coverage time of strategy f_D^m, i.e., the first time when every points in V is sweep-scanned, actually equals to the time when the last monitor, say M_j, finishes scanning the last point in its own subset V_j. Consequently, T_D^m equals to the maximum of m i.i.d. r.v.s, that is,

$$
T_D^m = \max_{M_i \in \mathcal{M}} \{T_{M_i}\}, \tag{5.38}
$$

which can be upper-bounded using the technique in [7, Eq.(3)],

$$
\begin{aligned}
\mathbb{E}(T_D) &\leq \max_{M_i \in \mathcal{M}} \{\mathbb{E}(T_{M_i})\} + \sqrt{(m-1)Var(T_{M_i})} \\
&= \mathbb{E}(T_{M_i}) + \sqrt{(m-1)Var(T_{M_i})}.
\end{aligned}
\tag{5.39}
$$

Plugging Eq. (5.37) into Eq. (5.39) yields the upper-bound. □

5.5.2.3 Numerical Validation

Figures 5.16 and 5.17 show the expected coverage time of I-strategy ($\mathbb{E}(T_I^m)$, blue '○' markers) and D-strategy ($\mathbb{E}(T_D^m)$, red '×' markers) with respect to $m \in [1, 10]$, number of monitors, and $n = |V| \in [50, 500]$, size of the assignment space, respectively. Numerical results of T_I and T_D are shown in dots, while their mean and standard deviation are shown with markers and bracketed bars. The case of four monitors ($m = 4$) is zoomed in the inner box of Fig. 5.16, and the distribution of coverage time (for $n = 200$ assignment points) is shown in the inner box of Fig. 5.17. We draw the following observations from the numerical validation:

1. Theorem 5.3 (red dashed lines in Figs. 5.16 and 5.17) is a tight bound on the coverage time of D-strategy when m is small.
2. Though not a tight bound, Theorem 5.2 (blue dotted lines in Figs. 5.16 and 5.17) accurately describes the $O(\frac{n}{m}\ln n)$ scaling behavior.
3. I-strategy and D-strategy have very close coverage time performances, not only in the mean sense, but also in distribution, as shown in the inner boxes of

Fig. 5.16 The expected coverage time of both the I- ($\mathbb{E}(T_I)$) and D-strategy ($\mathbb{E}(T_D)$) are scaling as $O(\frac{1}{m})$ over $m = |\mathcal{M}|$ with $n = 500$

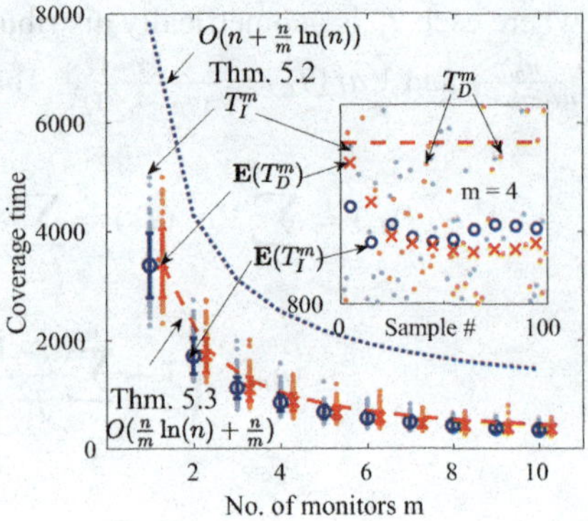

Fig. 5.17 The expected coverage time of both the I- ($\mathbb{E}(T_I)$) and D-strategy ($\mathbb{E}(T_D)$) are scaling as $O(n \ln n)$ over $n = |V|$ with $m = 5$

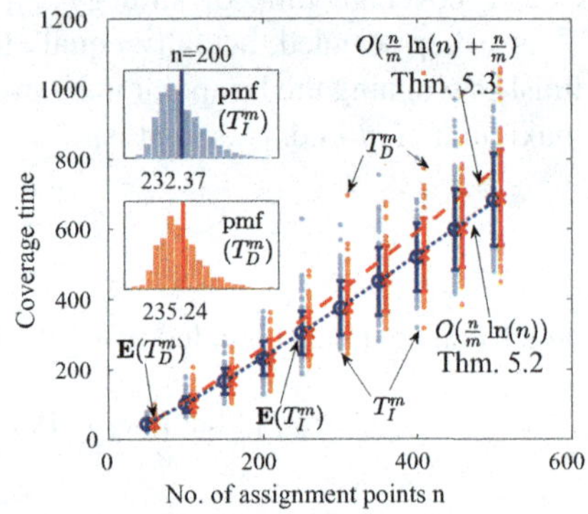

Fig. 5.17, which implies that the more demanding I-strategy (in terms of level of coordination and switching capability of monitors) can be safely substituted by the distributed D-strategy, with the same guaranteed performance in sweep-coverage.

4. Both expected coverage times can be described as $O(\frac{n}{m} \ln n)$ (blue dotted line in Fig. 5.17), indicating that this scaling law can be used to predict the number of monitors needed to reach the coverage goal for SAS over a fixed spectrum range and geographical region, when the resolution of surveillance result is determined by the $q(\delta)$-monitoring power.

Remark 5.3 Note that unit of the coverage time is time step, whose length can be evaluated when parameters in Table 5.4 are determined in a real-world scenario, according to the monitoring power of SAS monitors. We do not incorporate specific units in the analysis or simulation, because our main objective is to find the scaling behavior of strategy performance with respect to the problem size (in this case n) and the amount of SAS resource (in this case m), such that generic design guidelines can be obtained for different SAS application scenarios.

Compared with the deterministic strategy f_S, which achieves a $\frac{n}{m}$ coverage time, randomized strategies seem to be at disadvantage in fulfilling the coverage goal, but as we will show in the following subsection, they are favorable in culprit detection.

5.5.3 *Bounded Detection Time of Adversarial Culprits*

As briefly analyzed in the example (Fig. 5.15), the advantage of adversarial culprits over deterministic strategies will be lost, when facing monitors running randomized strategies f_I and f_D. The reason behind this is that, culprits' prior knowledge on visiting probabilities is compromised by the uniform probability distribution in f_I and f_D. Consequently, randomized strategies do not suffer from the 'wandering hole' problem, that is, the detection time is bounded.

Theorem 5.4 *Under strategy f_I^m and f_D^m, the expected detection time $\mathbb{E}(\tau_R(f_*))$ of an adversarial culprit R_a is upper-bounded, if the detection probability is strictly positive, i.e., $q \geq q^* > 0$. Particularly,*

$$\mathbb{E}\big(\tau_a(f_I^m)\big) = \Big[1 - (1 - \frac{q}{n})^m\Big]^{-1}, \tag{5.40}$$

$$\mathbb{E}(\tau_a(f_D^m)) = \frac{n}{qm}. \tag{5.41}$$

Proof Consider m monitors over the assignment space V of size $n = |V|$, each has $q(\delta)$-monitoring power. That is to say, if the spectra-location distance between a monitor M_i and a culprit R_a is less than δ, the probability that this particular culprit R_a will be detected by M_i during one time step with probability $q \geq q^*$. To I- and D-strategy, an adversarial culprit R_a is equivalent to a mobile persistent culprit R_{md} (see Table 5.2) in terms of detection time, because from the perspective of R_a, the probability of any assignment point being visited in the next time step is the same.

First we discuss the I-strategy f_I^m. The detection time is actually the first meeting time between the culprit and any of the m monitors, while both the culprit and monitors randomly walk on the complete graph K_n. By Eq. (5.29), detection time $\tau_R(f_I^m)$ is geometrically distributed with parameter $p_I = 1 - (1 - \frac{q}{n})^m$, which equals to the probability that R is caught by at least one of the m monitors in a single time step. Therefore,

$$\mathbb{E}\big(\tau_a(f_I^m)\big) = \frac{1}{p_I} = \Big[1 - (1 - \frac{q}{n})^m\Big]^{-1}. \tag{5.42}$$

Then given that the detecting probability/reliability q is lower-bounded by a positive constant $q^* > 0$, Eq. (5.40) can be upper-bounded by

$$\mathbb{E}(\tau_a(f_I^m)) \le \frac{1}{1 - \left[(1 - \frac{q^*}{n})^{\frac{n}{q^*}}\right]^{\frac{q^* m}{n}}} \le \frac{1}{1 - e^{-\frac{q^* m}{n}}}. \tag{5.43}$$

For the D-strategy f_D^m, each monitor switches independently, and occupies a different point in V during each time step. Therefore, the probability that culprit R_a is detected by any of the m monitors during t is $p_D = q\frac{m}{n}$, and the corresponding expected detection time is also upper-bounded:

$$\mathbb{E}(\tau_a(f_D^m)) = \frac{1}{p_D} = \frac{n}{qm} \le \frac{n}{q^* m}. \tag{5.44}$$

Therefore, any adversarial culprit R_a will be detected in finite time, given that detection probability $q \ge q^* > 0$. □

Remark 5.4 Theoretically, it is possible that detecting probability q is minimal, due to the large radius δ in the $q(\delta)$-monitoring power, so the resulting detection time $\mathbb{E}(\tau_a(f_I^m))$ and $\mathbb{E}(\tau_a(f_D^m))$ tend to infinity. However, this can be easily fixed if the radius parameter δ is adjusted in the space-tessellation step, so that q is boosted to an acceptable level. Consequently, we conclude that randomized monitoring strategies (f_I and f_D) do not suffer from the 'wandering hole' problem.

To validate the advantage of randomized strategies against adversarial culprit, i.e., Theorem 5.4, culprit detection is simulated under the same spectra-location space setting as the sweep-coverage validation. Detection time samples are shown as light-blue (I-strategy) and light-red (D-strategy) dots in Figs. 5.18 and 5.19, which corresponds to the $q = 1$ case (perfect detection), and Fig. 5.20, which corresponds to the imperfect detection ($q = 0.8 < 1$) case. We highlight the following observations.

1. Not only are the detection time of I and D-strategy bounded (and hence no 'wandering hole' problem), the exact values of their expectations can be obtained with Eqs. (5.40) and (5.41), once the number of monitors m and size of the assignment space n are fixed.

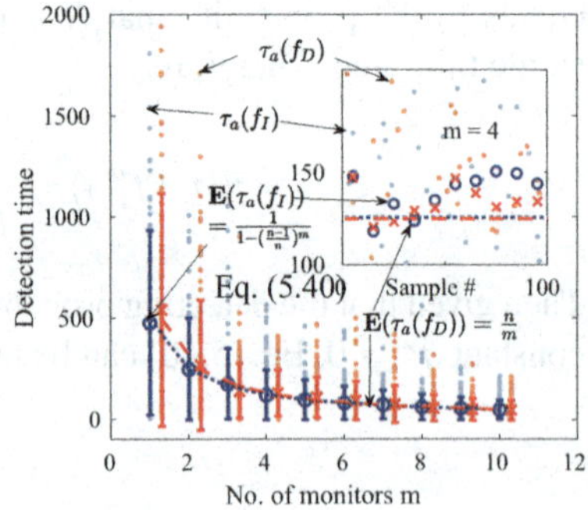

Fig. 5.18 The expected detection time of an adversarial culprit in the assignment space of size n scales as $O(\frac{1}{m})$ over $m = |M|$ with $n = 500$, under both the I and the D-strategy with full reliabiltiy $q = 1$

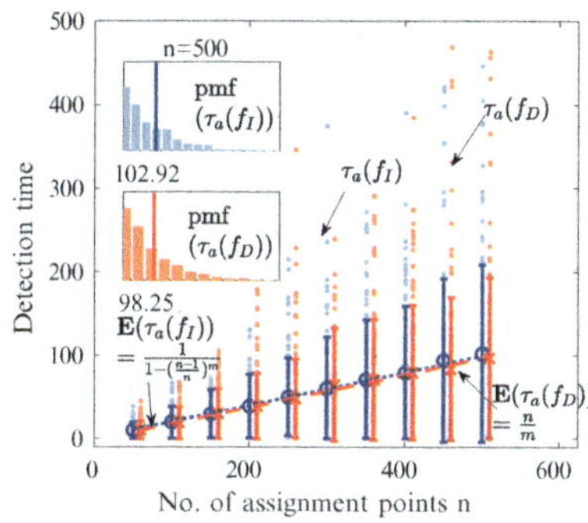

Fig. 5.19 The expected detection time of an adversarial culprit in the assignment space of size n scales as $O(n)$ over $n = |V|$ with $m = 5$, under both the I and the D-strategy with full reliability $q = 1$

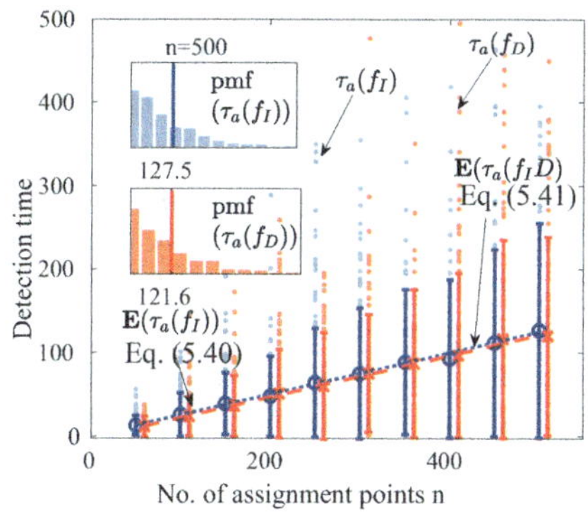

Fig. 5.20 Expected detection time of culprits under imperfect detection (reliability $q = 0.8$) is also well captured by Eqs. (5.40) and (5.41)

2. Bounds in Eqs. (5.40) and (5.41) also hold for the imperfect detection case, that is, detecting probability $q < 1$, as shown in Fig. 5.20.
3. The detection performance of the I- and D-strategy are fairly close, which indicates that D-strategy can be a good distributed alternative to the I-strategy.

In addition, the $O(\frac{1}{m})$ scaling behavior in both coverage (Fig. 5.16) and detection (Fig. 5.18) time, indicates a linear 'speed-up' in SAS performance, when multiple monitors are employed in the randomized strategies. This behavior implies, as same as in the deterministic strategies, increasing the number of monitors (m) is an efficient performance-boosting measure. In addition, the bounds on detection time (or rather, accurate results in Eqs. (5.40) and (5.41)) add to the predictability of randomized strategies, which can be fairly useful in the design stage of a SAS system, e.g., estimating the number of monitors needed for culprit detection.

5.6 SAS with Limited Switching Capacities

In this section, we discuss the SAS process of m independent α_M-monitors (with detection probability $q = 1$) and an α_R-culprit on the assignment space V, in which the switching capacity of the monitors (α_M) and the culprit (α_R) are limited, i.e., finite. Constrained switching capacity ($\alpha_M < \infty$) applies to the fully-distributed crowdsourcing SAS scenario, in which a switching is only possible if the two devices are within each other's communication range. Such a capacity limit can also be viewed as a binary quantification of the switching cost, in the sense that it cuts off any edge (v_i, v_j) (in a complete graph), whose associated switching cost $\gamma(v_i, v_j)$ exceeds a threshold, to obtain the monitoring/exploiting subgraphs. Analysis of a SAS process under switching capacity constraint, or as we formulate it, a composite graph walk process on graph (G_M, G_R) that is not a complete graph, is challenging due to the following reasons: (1) theoretic analysis of random walks on general graphs is difficult, if at all possible, because existing mathematical tools are developed for graphs with special structures, e.g., the complete graph [2] and the expander graph [9]; (2) monitors and culprits switch/walk on different graphs when $\alpha_R \neq \alpha_M$, which is not addressed by existing research on graph walk, e.g., [10].

For the ease of discussion, we introduce *degree* $r_*(\cdot)$ of an assignment point in the monitoring subgraph G_M and the exploiting subgraph G_R. Let $r_{\alpha_M}(v)$ denote the degree of point $v \in V$ in G_M, under constraint α_M, and $r_{\alpha_R}(v)$ denote the degree of v in G_R, under constraint α_R. Graph G_M and G_R are both subgraphs of a complete graph (degree $r(v) = n - 1$ for every node v), which corresponds to the case of $\alpha_M = \alpha_R = \infty$ discussed in Sect. 5.5. When $\alpha_M = \alpha_R$, the culprit and the monitors can be viewed as walking on the same graph, i.e., $G_M = G_R$. However, for cases when monitors are more 'powerful' than the culprit ($\alpha_M > \alpha_R$), edges in G_R are strictly sparser, i.e., $E_R \subset E_M$, and vise versa.

5.6.1 *Performance Analysis Through Regular Graph Approximation*

Observe that assignment points (cell centers in the Kevin structure) in space V are quite 'structured', as shown in Fig. 5.5. As a result, subgraph G_M (respectively G_R) (e.g., the weaker monitors vs. powerful culprit case shown in Fig. 5.8) that build upon it is also 'structured', in the sense that degrees of most vertices in G_M (G_R) are roughly the same, except for the few near the boundary. Denote $r_M = \frac{1}{n}\sum_{i=1}^{n} r_{\alpha_M}(v_i)$ as the average degree of the monitoring subgraph G_M, and r_R as that of the exploiting subgraph G_R. We first approximate G_M and G_R as r_M-and r_R-regular graphs[15] (G_{r_M} and G_{r_R}), respectively, on which mathematical tools such as [2, 3, 10] come in handy in the performance analysis.

[15] Though reasonable, this average-degree-based approximation introduces a gap when determining SAS performance for both sweep-coverage and culprit detection, which is discussed in Sect. 5.6.2.

5.6.1.1 Coverage Time $T_{r_M}^m$

Let $T_{r_M}^m$ denote the coverage time of m independent monitors on r_M-regular graph G_{r_M}, to differentiate from the actual coverage time T_I^m on the original monitoring subgraph G_M. On regular graph G_{r_M}, asymptotic bounds for $\mathbb{E}(T_{r_M}^m)$ have been studied by multiple researchers. Among these, Alon et al. [3] proved $\mathbb{E}(T_{r_M}^m) \sim \Theta(\frac{n \ln n}{m})$, as $n \to \infty$; Cooper et al. [10] provided a similar but more accurate asymptotic result for random regular graphs, when the number of random walkers is not large, i.e., $m = o(\frac{n}{\ln^2 n})$. It is shown in [9] that a uniformly chosen r−regular ($r \geq 3$) graph G_r is 'nice' with high probability (tending to one as $n \to \infty$), such that the expected coverage time $T_{r_M}^m$ follows from [10, Thm. 2], that is,

$$\mathbb{E}(T_{r_M}^m) \sim \frac{r_M - 1}{r_M - 2} \frac{n \ln n}{m}. \tag{5.45}$$

Under an I-strategy, each monitor switches to any assignment point within its switching capacity uniformly at random, resulting in a uniform stationary distribution $\pi_v = \frac{1}{n}$ over the assignment space V. In other words, each assignment point is visited roughly the same number of times as time proceeds, so we say the spectra-location space X is 'evenly' covered. Nevertheless, if a certain region (subspace of X) needs special attention, e.g., due to higher presence of misbehavior, the probability to switch to a target assignment point in the switching capacity can be adjusted, such that a desired (possibly non-uniform) stationary distribution over V can be achieved through well-articulated algorithms, e.g., the Metropolis-Hasting algorithm.

5.6.1.2 Detection Time $\tau_R(r_R, r_M)$

Unlike coverage time, for which existing research leads to direct solution, there is no proper mathematical tool to directly address the culprit detection problem, in which the monitors and the culprit may walk on *different* graphs, due to their different switching capacity limits. So we address the weaker monitors vs. powerful culprit ($\alpha_M < \alpha_R$) case (and the opposite case $\alpha_M > \alpha_R$) in this subsection.

Let $\tau_R(r_R, r_M)$ denote the detection time of a culprit R (walking on the r_R-regular graph G_{r_R}), by m-monitors (walking on the r_M-regular graph G_{r_M}). There are two possible cases:

Case 1. *Same switching capacity* ($r_R = r_M = r$). Monitors $\mathcal{M}$ and the culprit R have the same switching capacity, such that $r_R = r_M = r$, and $G_{r_R} = G_{r_m} = G_r$, i.e., both monitors and culprit R walk on an r-regular graph G_r. Applying the predator-and-prey model [10, Theorem 3], the expected detection time of culprit R can be asymptotically bounded as

$$\mathbb{E}(\tau_R(r, r)) \sim \frac{r - 1}{r - 2} \cdot \frac{n}{m}. \tag{5.46}$$

Case 2. *Different switching capacities* ($r_R \neq r_M$). Monitors $\mathcal{M}$ and the culprit R walk on different graphs, i.e., $r_R \neq r_M$ such that $G_{r_R} \neq G_{r_m}$. We obtain the upper bound of $\mathbb{E}(\tau_R(r_R, r_M))$ by considering a composite random walk.

Proposition 5.2 *Let* $K = \frac{(n-1)!}{(n-m-1)!}$. *Under an I-strategy with m monitors, the expected detection time of a culprit on graph* (G_{r_M}, G_{r_R}) *is upper bounded ass*

$$\mathbb{E}(\tau_R(r_R, r_M)) \leq 1 + \frac{K}{n^m}(4K^2 - 1). \tag{5.47}$$

Proof Without loss of generality, we number the monitors in $\mathcal{M}$ as $\{1, 2, \ldots, m\}$. Let $\bar{\mathcal{M}} = \mathcal{M} \cup \{m+1\}$ denote a set of walkers, where the first m walkers correspond to monitors, who walk on the monitoring subgraph G_M, and walker $m+1$ refers to the culprit R, who walks on the exploiting subgraph G_R.

Consider a (single-walker) random walk $\{Z_t\}_{t \in \mathcal{T}}$ on graph $H = (V_H, E_H)$, where each vertex $\mathbf{v}$ is a vector consisting $m+1$ elements, that is, $V_H = \{\mathbf{v} = (v_1, v_2, \ldots, v_m, v_{m+1}) \mid v_i \in V,\ \forall i \in \bar{\mathcal{M}}\}$. Edges of graph H are constructed as follows: edge $(\mathbf{v}, \mathbf{w}) \in E_H$ exists, if and only if: (1) $(v_i, w_i) \in E_M,\ \forall i \in \mathcal{M}$; and (2) $(v_{m+1}, w_{m+1}) \in E_R$. Graph H is a regular graph, since for every vertex $\mathbf{v} \in V_H$, the number of neighbors, or the degree of vertex $\mathbf{v}$, $d_H(\mathbf{v}) = r_M^m \cdot r_R$, given that (G_M, G_R) is approximated by the regular composite graph (G_{r_M}, G_{r_R}).

Now the SAS process, or more specifically, the culprit detection process, can be described by the single-walker random walk $\{Z_t\}_{t \in \mathcal{T}}$ on graph H, thanks to its construction. Let set

$$B := \{\mathbf{v} \in V_H \mid \exists\, i \in \mathcal{M}\ s.t.\ v_i = v_{m+1}\}$$

denote the set of vertices where the culprit (walker $m+1$) is detected by one of the m monitors (co-locates with one of the m walkers in $\mathcal{M}$). Then the detection time of culprit R, is the first hitting time of set B by walk Z_t, that is,

$$\tau_R^{\mathbf{v}}(r_R, r_M) = \min_{t \in \mathcal{T}} \{t \mid Z_t \in B\}, \tag{5.48}$$

provided that walk Z_t started from initial position $Z_1 = \mathbf{v}$. Since both the monitors $\mathcal{M}$ and culprit R starts uniformly at random in the assignment space V at time $t = 1$, the probability that walk Z_t starts from vertex $\mathbf{v}$ can be calculated by $\mathbb{P}(Z_1 = \mathbf{v}) = \frac{1}{|V_H|} = n^{-(m+1)}$. With respect to set B, there are two possible cases:

Case 1. If walk Z_t starts from set B, that is, $Z_1 \in B$, the culprit R is immediately identified. This event happens with probability

$$\mathbb{P}\left(\tau_R^{\mathbf{v}}(r_R, r_M) = 1\right) = \mathbb{P}(Z_1 \in B) = \frac{|B|}{|V_H|} = 1 - \frac{n!}{(n-m-1)!} = 1 - nq. \tag{5.49}$$

Case 2. Otherwise (with probability $1 - \mathbb{P}\left(\tau_R^{\mathbf{v}}(r_R, r_M) = 1\right)$), the initial position $Z_1 \in V_H \setminus B$, with probability nq. In this case, since H is a regular, the following inequality follows from [2, Proposition 6.16]:

$$\mathbb{E}\left(\tau_R^{\mathbf{v}}(r_R, r_M) \mid \mathbf{v} \notin B\right) \leq 4|V_H \setminus B|^2. \tag{5.50}$$

Then the upper bound in Eq. (5.47) can be obtained by combing the two cases.

□

Proposition 5.2 holds for every n, but when n is large, it is not easy to obtain K and n^m in Eq. (5.47). For this case, we have the following scaling law result.

Corollary 5.1 *The expected detection time of a culprit for a SAS process under the I-strategy on graph (G_{r_R}, G_{r_M}) satisfies*

$$\mathbb{E}\left(\tau_R(r_R, r_M)\right) = \Theta(\frac{n}{m}), \tag{5.51}$$

where the exploiting subgraph G_{r_R} and the monitoring subgrpah G_{r_M} are r_R- and r_M-regular graphs, respectively.

Proof Let $r_1 = \min\{r_R, r_M\}$ and $r_2 = \max\{r_R, r_M\}$, then

$$\mathbb{E}\left(\tau_R(r_2, r_2)\right) \leq \mathbb{E}\left(\tau_R(r_R, r_M)\right) \leq \mathbb{E}\left(\tau_R(r_1, r_1)\right). \tag{5.52}$$

Also, from results of detection time on regular graphs, we have $\mathbb{E}\left(\tau_R(r_2, r_2)\right) \sim \frac{r_2-1}{m(r_2-2)}n = \Omega(\frac{n}{m})$ and $\mathbb{E}\left(\tau_R(r_1, r_1)\right) \sim \frac{r_1-1}{m(r_1-2)}n = O(\frac{n}{m})$. Then it easily follows that $\mathbb{E}\left(\tau_R(r_R, r_M)\right) = \Theta(\frac{n}{m})$. □

Compared to the SAS scenarios with unlimited switching capacity ($\alpha_M = \infty$) discussed in Sect. 5.5, the scaling laws in this regular graph approximation (coverage time Eq. (5.45), detection time Eqs. (5.46) and (5.51)) differ only by a degree-determining constant, which is less than or equal to 2. Consequently, we expect that, upon the imposed switching capacity limit, scaling laws of both performance metrics over m and n to remain the same compared to that under no switching capacity limit.

5.6.2 Gap Between (G_M, G_R) and Approximation (G_{r_M}, G_{r_R})

For Eqs. (5.45) and (5.46) to hold, a regular graph needs to be 'nice' [9, pp. 733]. It is also shown [9] that a large (n large) r-regular graph G_r randomly selected from all r-regular graphs $\mathcal{G}_r$, is *almost-Ramanujan* with high probability, that is, the largest eigenvalue $\lambda_0(G_r)$ and the second largest eigenvalue $\lambda_1(G_r)$ of graph G_r's adjacency matrix satisfy

$$\lambda_1(G_r) \leq 2\sqrt{\lambda_0(G_r) - 1} + \epsilon, \tag{5.53}$$

where the $\lambda_0(G_r) = r$, as G_r is r-regular.

However, Eq. (5.53) does not necessarily hold for the actual composite graph (G_R, G_M). For instance, the monitoring subgraph G_M presented in Fig. 5.8 corresponds to $\alpha_M = 5$, and it has $\lambda_0(G_M) = 18.415$ and $\lambda_1(G_M) = 16.475$, certainly violating the eigenvalue gap criterion in Eq. (5.53), while a randomly generated graph G_{r_M} ($r_M = 17$), with the same average degree as graph G_M, has $\lambda_0(G_{r_M}) = 17$ and $\lambda_1(G_{r_M}) = 7.633$ satisfying the criterion.

This gap in the graph expansion property does not allow direct application of the scaling law (described by Eqs. (5.45) and (5.46)) to the composite graph (G_M, G_R), which is induced by switching capacity limit α_M and α_R, even though (G_M, G_R) have the same average degree as its approximation (G_{r_M}, G_{r_R}) by construction. Therefore, we employ simulation to see if the approximation is valid.

5.6.3 Numerical Results

We validate the regular approximation in the same assignment space V detailed in Table 5.4. Simulation results (dots) and bounds (dashed and dotted lines) of the coverage time and detection time, under the I-strategy with different switching capacities, are shown in Figs. 5.21 and 5.22, respectively. The powerful monitors case ($\alpha_M = 10$, corresponding to $r_M = 19$, and $\alpha_R = 5$, corresponding to $r_R = 86$) is marked in blue, whose mean is shown by blue '○' marker, while the powerful culprit case ($\alpha_M = 5$, $\alpha_R = 10$) is marked in red, whose mean is shown by red '×' marker. In Fig. 5.21, the lower bound of the coverage time (black dotted line) is obtained by setting r_M to ∞ in Eq. (5.45).

We highlight the following observations from these results.

1. For the coverage time in Fig. 5.21, we observe as we anticipated. Specifically,
   (a) The coverage time by a set of weaker monitor (red '×' markers, switching capacity limit $\alpha = 5$) is slightly longer than that by a set of more powerful monitors (blue '○' markers, $\alpha_M = 10$).

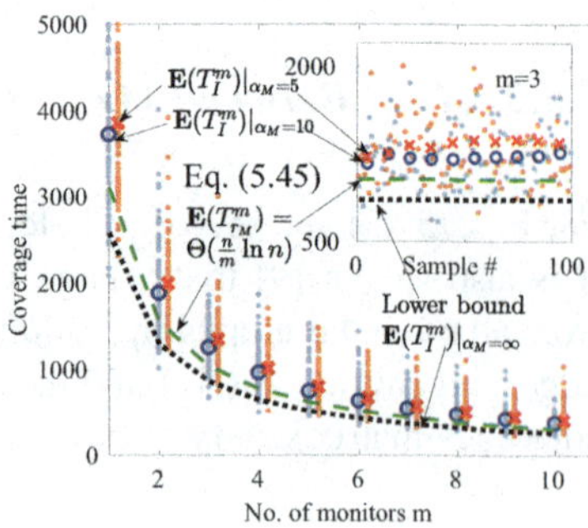

Fig. 5.21 The expected coverage time $\mathbb{E}(T_I^m) = \Theta(\frac{n}{m}\ln n)$ for SAS processes with switching capacity limits, in which bounds are derived with regular graph approximation

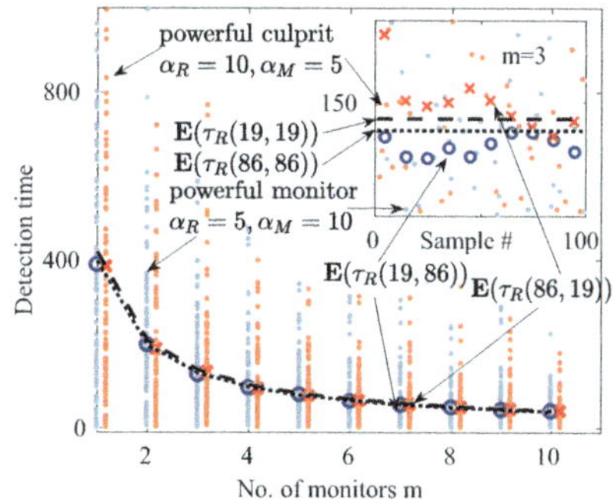

Fig. 5.22 The expected detection time $\mathbb{E}(\tau_a(f_I^m)) = \Theta(\frac{n}{m})$ for SAS processes with switching capacity limits, in which bounds are derived with regular graph approximation

(b) The $\Theta(\frac{n}{m}\ln n)$ scaling law of the expected coverage time $\mathbb{E}(T_I^m)$ over m is well captured, despite the switching capacity limit α_M.

2. For the detection time in Fig. 5.22, these results also echo with our analysis.

(a) As predicted by Corollary 5.1, the expected detection time $\mathbb{E}(\tau_R(f_I^m))|_{\alpha_M=5}$ is not lengthened much, compared to the strengthened case $\alpha_M = 10$, as opposed to an intuitive anticipation. This indicates that both the time and range aspects of the switching capacity do not impact the expected detection time much.

(b) Both the upper and lower bounds of the expected detection time are tight, if not precise ($\mathbb{E}(\tau_R(f_I^M)) \simeq \mathbb{E}(\tau_R(r_R, r_M))$ for $m \in [1, 10]$).

3. Some common effects apply to both the coverage time and detection time.

(a) Even though the switching capacity of monitors (α_M) and that of the culprit (α_R) differ considerably in value for the two simulation cases, the mean coverage and detection time (round and '×' markers in both Figs. 5.21 and 5.22) are pretty close. The reason behind this is similar to what is revealed in Lemma 5.1, i.e., the more 'mobile' (either monitors or culprits), the more 'visible' to spectrum monitors.

(b) Though (G_M, G_R) are not regular graphs, bounds derived for their regular graph approximation (G_{r_M}, G_{r_R}) (dash and dotted black lines) apply smoothly to both the coverage time and the detection time, in the sense that the scaling laws are well-captured in both figures.

Comparing the unlimited switching capacity cases (shown in Figs. 5.16 and 5.18) with the limited cases (shown in Figs. 5.21 and 5.22), we observe that the capability limit α_M becomes less influential as the number of monitors m increases, and does not change the scaling behavior over m. The reason behind this is that α_M is sufficiently large, so that the quantity $\frac{r_M-1}{r_M-2}$ in Eq. (5.45) comes close to 1. With extensive simulation, we find that $\mathbb{E}(T_I^m)$ and $\mathbb{E}(\tau_R(f_I^m))$ on the real composite graph (G_M, G_R) actually follow the $\Theta(\frac{n\ln n}{m})$ and $\Theta(\frac{n}{m})$ scaling law described in Eqs. (5.45) and (5.46)). We speculate the reason behind this is that both subgraphs (G_M and G_R) are well-connected in degree sense, and the variation in node degree is small, such that G_M and G_R are 'regular' enough. On the other hand, this observation makes us wonder whether the requirement of being 'nice' is necessary in achieving the $\Theta(\frac{n \ln n}{m})$ and $\Theta(\frac{n}{m})$ scaling law.

5.7 Summary

From the impact aspect of mobile data, in this chapter, we examine spectrum activity surveillance (SAS) for DSA-enabled wireless systems, which are envisioned as the monumental delivery network for many data services. We identify the two main objectives of SAS, as sweep-coverage of spectra-location space, and detection of simple/adversarial spectrum culprits, for which we define quantitative metrics, such that any SAS monitor deployment strategy can be fairly evaluated. To accurately describe SAS processes, we introduce a 3-D model that incorporates spectra, temporal and geographical domains, and captures the locality of different spectrum activities. Based on this model, we formulate a SAS process into a graph walk, by space tessellation and pruning with the switching capacity limits of monitors. As an application of the proposed model, we present a deterministic strategy to achieve low switching cost, and randomized strategies to quickly catch adversarial culprits. Efficacy of these strategies is theoretically analyzed and validated through simulations, revealing how the size of a SAS system and the amount of monitoring resource affect their performance. We hope these results contribute to the knowledge of spectrum surveillance, and benefit the design of DSA-enabled wireless systems.

References

1. 3GPP: NR; User Equipment (UE) radio transmission and reception; Part 3: Range 1 and Range 2 Interworking operation with other radios. Technical Specification (TS) 38.101-3, 3rd Generation Partnership Project (3GPP), version 15.4.0 (2019). http://www.3gpp.org/\discretionary-DynaReport/\discretionary-38101-3.htm
2. Aldous, D., Fill, J.A.: Reversible markov chains and random walks on graphs (2002). Unfinished monograph, recompiled 2014, available at https://www.stat.berkeley.edu/~aldous/RWG/book.pdf
3. Alon, N., Avin, C., Koucky, M., Kozma, G., Lotker, Z., Tuttle, M.R.: Many random walks are faster than one. Combin. Prob. Comput. **20**(4), 481–502 (2011). https://doi.org/10.1017/S0963548311000125
4. Baldini, G., Steri, G.: A survey of techniques for the identification of mobile phones using the physical fingerprints of the built-in components. IEEE Commun. Surv. Tutor. **19**(3), 1761–1789 (2017). https://doi.org/10.1109/COMST.2017.2694487
5. Bektas, T.: The multiple traveling salesman problem: an overview of formulations and solution procedures. Omega **34**(3) (2006). http://dx.doi.org/10.1016/j.omega.2004.10.004. http://www.sciencedirect.com/science/article/pii/S0305048304001550
6. den Bergh, B.V., Giustiniano, D., Cordobés, H., Fuchs, M., Calvo-Palomino, R., Pollin, S., Rajendran, S., Lenders, V.: Electrosense: Crowdsourcing spectrum monitoring. In: 2017 IEEE International Symposium on Dynamic Spectrum Access Networks (DySPAN), pp. 1–2 (2017). https://doi.org/10.1109/DySPAN.2017.7920766

7. Bertsimas, D., Natarajan, K., Teo, C.P.: Tight bounds on expected order statistics. Probab. Eng. Inf. Sci. **20**(4), 667–686 (2006). https://doi.org/10.1017/S0269964806060414
8. Chen, Y., Ding, M., Lopez-Perez, D., Li, J., Lin, Z., Vucetic, B.: Dynamic reuse of unlicensed spectrum: An inter-working of lte and wifi. IEEE Wirel. Commun. **24**(5), 52–59 (2017). https://doi.org/10.1109/MWC.2017.1700088
9. Cooper, C., Frieze, A.: The cover time of random regular graphs. SIAM J. Discrete Math. **18**(4), 728–740 (2005). https://doi.org/10.1137/S0895480103428478
10. Cooper, C., Frieze, A., Radzik, T.: Multiple Random Walks and Interacting Particle Systems, pp. 399–410. Springer Berlin Heidelberg, Berlin/Heidelberg (2009). https://doi.org/10.1007/978-3-642-02930-1_33
11. Ding, Z., Lei, X., Karagiannidis, G.K., Schober, R., Yuan, J., Bhargava, V.K.: A survey on non-orthogonal multiple access for 5g networks: Research challenges and future trends. IEEE J. Sel. Areas Commun. **35**(10), 2181–2195 (2017). https://doi.org/10.1109/JSAC.2017.2725519
12. Google: Google spectrum database: Google earth visualization of available tv white space spectrum. https://www.google.com/get/spectrumdatabase/ (2013). Accessed: Dec 18, 2017
13. Hatfield, D.N., Claudy, L., Gorenberg, M., Gurney, D., Lapin, G., Markwalter, B., Mendenhall, G., de Vries, P., Roberson, D.: Introduction to interference resolution, enforcement and radio noise. Technical Report, FCC 14-31, FCC Technological Advisory Council (2014)
14. Holland, O., Bogucka, H., Medeisis, A.: Practical Mechanisms Supporting Spectrum Sharing, pp. 141. Wiley Telecom, Hoboken (2015). https://doi.org/10.1002/9781119057246.part2. http://ieeexplore.ieee.org/xpl/articleDetails.jsp?arnumber=8043315
15. Höyhtyä, M., Mämmelä, A., Eskola, M., Matinmikko, M., Kalliovaara, J., Ojaniemi, J., Suutala, J., Ekman, R., Bacchus, R., Roberson, D.: Spectrum occupancy measurements: a survey and use of interference maps. IEEE Commun. Surv. Tutor. **18**(4), 2386–2414 (2016). https://doi.org/10.1109/COMST.2016.2559525
16. Jain, M., Choi, J.I., Kim, T., Bharadia, D., Seth, S., Srinivasan, K., Levis, P., Katti, S., Sinha, P.: Practical, real-time, full duplex wireless. In: Proceedings of the 17th Annual International Conference on Mobile Computing and Networking, MobiCom '11, pp. 301–312. ACM, New York (2011). https://doi.org/10.1145/2030613.2030647. http://doi.acm.org/10.1145/2030613.2030647
17. Jin, X., Sun, J., Zhang, R., Zhang, Y., Zhang, C.: Specguard: Spectrum misuse detection in dynamic spectrum access systems. In: 2015 IEEE Conference on Computer Communications (INFOCOM), pp. 172–180 (2015). https://doi.org/10.1109/INFOCOM.2015.7218380
18. Li, M., Yang, D., Lin, J., Li, M., Tang, J.: Specwatch: adversarial spectrum usage monitoring in crns with unknown statistics. In: IEEE INFOCOM 2016 - The 35th Annual IEEE International Conference on Computer Communications, pp. 1–9 (2016). https://doi.org/10.1109/INFOCOM.2016.7524555
19. Liang, C., Yu, F.R.: Wireless network virtualization: a survey, some research issues and challenges. IEEE Commun. Surv. Tutor. **17**(1), 358–380 (2015). https://doi.org/10.1109/COMST.2014.2352118
20. Liu, S., Greenstein, L.J., Trappe, W., Chen, Y.: Detecting anomalous spectrum usage in dynamic spectrum access networks. Ad Hoc Netw. **10**(5), 831 – 844 (2012). http://dx.doi.org/10.1016/j.adhoc.2011.04.003. http://www.sciencedirect.com/science/article/pii/S1570870511000801. Special Issue on Cognitive Radio Ad Hoc Networks
21. Liyanage, M., Gurtov, A., Ylianttila, M.: Leveraging SDN for the 5G Networks, pp. 61–80. Wiley Telecom, Hoboken (2015). https://doi.org/10.1002/9781118900253.ch5. http://ieeexplore.ieee.org/xpl/articleDetails.jsp?arnumber=8044821
22. Microsoft: White spaces database (2015). http://whitespaces.microsoftspectrum.com/. Accessed: Feb 11, 2018
23. Nika, A., Zhang, Z., Zhou, X., Zhao, B.Y., Zheng, H.: Towards commoditized real-time spectrum monitoring. In: Proceedings of the 1st ACM Workshop on Hot Topics in Wireless, HotWireless '14, pp. 25–30. ACM, New York (2014). https://doi.org/10.1145/2643614.2643615. http://doi.acm.org/10.1145/2643614.2643615

24. Pan, M.J.L., Clancy, T.C., McGwier, R.W.: A machine learning approach for dynamic spectrum access radio identification. In: 2014 IEEE Global Communications Conference, pp. 1041–1046 (2014). https://doi.org/10.1109/GLOCOM.2014.7036946
25. Pfammatter, D., Giustiniano, D., Lenders, V.: A software-defined sensor architecture for large-scale wideband spectrum monitoring. In: Proceedings of the 14th International Conference on Information Processing in Sensor Networks, IPSN '15, pp. 71–82. ACM, New York (2015). https://doi.org/10.1145/2737095.2737119. http://doi.acm.org/10.1145/2737095.2737119
26. Rohde & Schwarz: Rohde & Schwarz® fsh4/8/13/20 spectrum analyzer operation manual (2016). https://www.rohde-schwarz.com/us/manual/r-s-fsh4-8-13-20-operating-manual-manuals-gb1_78701-29159.html. Accessed: Feb 21, 2018
27. Shin, D.H., Bagchi, S.: Optimal monitoring in multi-channel multi-radio wireless mesh networks. In: Proceedings of the Tenth ACM International Symposium on Mobile Ad Hoc Networking and Computing, MobiHoc '09, pp. 229–238. ACM, New York (2009). https://doi.org/10.1145/1530748.1530780. http://doi.acm.org/10.1145/1530748.1530780
28. Sundar, K., Rathinam, S.: Generalized multiple depot traveling salesmen problem–polyhedral study and exact algorithm. Comput. Oper. Res. **70**, 39 – 55 (2016). https://doi.org/10.1016/j.cor.2015.12.014. http://www.sciencedirect.com/science/article/pii/S0305054815003007
29. Tehrani, M.N., Uysal, M., Yanikomeroglu, H.: Device-to-device communication in 5g cellular networks: challenges, solutions, and future directions. IEEE Commun. Mag. **52**(5), 86–92 (2014). https://doi.org/10.1109/MCOM.2014.6815897
30. Thapaliya, A., Sengupta, S.: Understanding the feasibility of machine learning algorithms in a game theoretic environment for dynamic spectrum access. In: 2017 International Symposium on Performance Evaluation of Computer and Telecommunication Systems (SPECTS), pp. 1–8 (2017). https://doi.org/10.23919/SPECTS.2017.8046784
31. Thomson, S.W.: Lxiii. on the division of space with minimum partitional area. Philos. Mag. Series 5 **24**(151) (1887). https://doi.org/10.1080/14786448708628135. http://dx.doi.org/10.1080/14786448708628135
32. Voicu, A.M., Simić, L., Petrova, M.: Inter-technology coexistence in a spectrum commons: a case study of wi-fi and lte in the 5-ghz unlicensed band. IEEE J. Sel. Areas Commun. **34**(11), 3062–3077 (2016). https://doi.org/10.1109/JSAC.2016.2615246
33. Weaire, D., Phelan, R.: A counter-example to kelvin's conjecture on minimal surfaces. Philos. Mag. Lett. **69**(2), 107–110 (1994). https://doi.org/10.1080/09500839408241577. http://dx.doi.org/10.1080/09500839408241577
34. Yang, L., Zhang, Z., Zhao, B.Y., Kruegel, C., Zheng, H.: Enforcing dynamic spectrum access with spectrum permits. In: Proceedings of the Thirteenth ACM International Symposium on Mobile Ad Hoc Networking and Computing, MobiHoc '12, pp. 195–204. ACM, New York (2012). https://doi.org/10.1145/2248371.2248400. http://doi.acm.org/10.1145/2248371.2248400
35. Yoon, S., Li, L.E., Liew, S.C., Choudhury, R.R., Rhee, I., Tan, K.: Quicksense: fast and energy-efficient channel sensing for dynamic spectrum access networks. In: 2013 Proceedings IEEE INFOCOM, pp. 2247–2255 (2013). https://doi.org/10.1109/INFCOM.2013.6567028

Chapter 6
Conclusion and Future Directions

In previous chapters, we have presented our observations and results on mobile data dynamics in heterogeneous wireless networks, from four aspects: the cause, description, governing rule, and impact, respectively. These efforts are merely a small step toward the ultimate goal of provisioning for edge-native AI, or *Edge Intelligence (EI)*. In this chapter, we summarize our findings, then discuss challenges and open questions that are relevant and timely to this emerging research field.

6.1 Recap of Key Findings on Mobile Data Dynamics

This book discussed four aspects of mobile data in a top-down manner. We recapitulate our major findings on mobile data dynamics as follows.

We first studied the driving force of mobile data, that is, information dynamics, to answer *when* data start and stop moving in networks in Chap. 2. Specifically, we proposed a SIC epidemic model to study the conflicting information propagation phenomenon in networks, particularly the transient competition between the two during the evolution. Our analysis revealed the impact of network topology, propagation parameters, and the initial conditions, on the lifetime of the undesired information in a network. Upper bounds of the extinction time indicate a dominant impact of network topology, captured by the Cheeger constant, which determines the scaling laws of the lifetime over the network size. As an application of the model and results, we proposed practical and efficient information control measures, that is, injecting desired information into the network, and designed an inference algorithm to obtain the number of information adopters, before the conflicting information propagation process fully unfolds. Our findings quantitatively revealed the impact of network topology on the lifetime time of mobile data, which contributes to the design, management, and restoration of networked systems.

J. Wang et al., *Encountering Mobile Data Dynamics in Heterogeneous Wireless Networks*, https://doi.org/10.1007/978-3-031-62906-8_6

Then in Chap. 3, we studied *where* data can be accessed, during its dissemination process. Specifically, we examined data coverage, that is, locations where data are accessible, during the dissemination process in a heterogeneous wireless network. We defined data-strength to quantify data coverage as a numeric signal on a graph, which captures user mobility and geographical adjacency. Then, with graph signal processing (GSP), we observe high, decreasing, but bounded impact, of user mobility on the change of data coverage, indicating the prediction of data coverage is possible, with advanced GSP measures.

Onto the governing rule of mobile data, i.e., *how* data move and *what impact* mobile data causes, we investigated the task offloading process in the fog, which is a resource-constrained system residing at the network edge. We studied how multiple tasks, as special types of data blocks, move during offloading processes given the resource constraint, as well as the impact, i.e., resource demand, of mobile data on the provisioning networks. We proposed a gravity-based task offloading model, which captures the probabilistic offloading decisions based on various criteria, and defined device and network efforts with respect to data movements, to quantify the resource need for the massive task offloading procedures in the fog. With this model, we found that the time to complete individual tasks decreases with network size, while the total resource consumption by the system scales linearly with the network size. Our findings addressed the scalability issue of the fog paradigm, which is especially meaningful for large-scale fog applications, such as IoT.

Finally, we examined spectrum activity surveillance (SAS), to observe the *impact* of mobile data in Chap. 5. We analyzed SAS for DSA-enabled wireless systems, in order to observe the impact of mobile data on radio spectrum, which is a scarce resource at the network edge. We studied this problem in a 3-D space that incorporates spectra, temporal and geographical domains, and formulated the surveillance process into tractable graph walks. We defined two performance metrics, namely the coverage time and detection time, for spectrum monitor deployment strategies, such that any strategy can be fairly evaluated and compared. For the surveillance scenarios by dedicated spectrum monitors, and that by crowdsourcing spectrum monitors, we designed low-cost deterministic strategy for fast surveillance coverage, and effective randomized strategies for quick detection of spectrum culprits, respectively, offering useful design guidelines for large-scale spectrum surveillance systems.

In order to provide efficient, reliable, and accessible data services over future wireless networks, which is expected to host numerous heterogeneous devices, our study may be extended in the following directions.

Due to the open nature of current and future wireless systems, data are not only mobile in the geographical space domain, but also mutating in content, representation, and value, as they age. A natural follow-up question is, how does such mutation affect the mobility and accessibility of data? For example, a semantically complete data block can be fragmented and/or encrypted, when it is disseminated through the network. In this case, the data coverage, where this data block is accessible, becomes a process on a higher-dimensional structure that captures location, access capability, and data integrity, instead of the 2-D graph studied in Chap. 3. It would be very interesting to study these phenomena of data dissemination processes, for the benefit of data service provisioning.

Considering the exponentially growing size of wireless systems, powered by the proliferating wireless devices, the speed of network expansion (in size) may well surpass the speed of network capacity increase in a wireless system. For instance, in the fog paradigm, our work in Chap. 4 revealed that the growing number of offloading service participants will result in a linearly growing demand on network resources, in the form of radio spectrum, resource blocks, and so on. However, the total amount of network resource in the current wireless system is upper bounded, which indicates that the scale of data offloading service can not grow indefinitely. Consequently, it is highly desirable to obtain such limits, as an important characteristics of a wireless system, and a key input for designing data services. On the other hand, spectrum dynamics, as a result of mobile data, in turn limits the mobility of data, especially in densely populated networks, such as a fog system designed for IoT applications. Consequently, it will be equally necessary to further study the inverse of the impact problem addressed in Chap. 5, that is, what is the impact of spectrum dynamics on mobile data?

6.2 Emerging Trends on Edge Intelligence

As we step into the era of AI, the edge network is drawing more and more attention from researchers, as it is expected to host distributed AI [1, 2, 5], for the reduction of service latency, and the protection of data privacy. Thanks to the proliferation of IoT devices, fresh data, the indispensable fuel of AI models, are being generated at a staggering rate in the edge network. To cater for the stringent delay requirements of recent AI applications, such as autonomous driving [3], remote healthcare [4], and smart manufacturing [6], model training and inference tasks are gradually migrating to computing devices in the edge network. On the other hand, traditional model training that demands user data to be collected and compiled in centralized data centers raises great concerns for privacy leakage, so both users and researchers are leaning toward the privacy-preserving distributed learning schemes [5], such as federated learning. Such paradigm shift brings AI to the edge network, creating the emerging interdiscipline, Edge AI, or *Edge Intelligence (EI)* [7].

While early researches refer to EI as the paradigm of running AI algorithms locally on end devices, with data created on each individual device alone, a slightly different consensus has recently been reached by most scholars in this field, that is, EI is not restricted to devices edge network, but includes any possible computing paradigm that leverages the cloud-edge-things architecture. In fact, Zhou et al. proposed to rate EI as six levels [7], with different levels of participation from the cloud, edge, and individual devices. Aiming to unleash the full potential of edge big data, EI is expected to take AI applications to the next level, where computing, training, and inference are provided as pervasive services, open to a much broader range of users, including both large cooperate users, and individual users.

To this end, existing research efforts mainly focus on the design of computing paradigms, adaptation of existing machine learning models, distributed learning algorithms, and deployment frameworks. These topics are all relevant to mobile data dynamics discussed in this book, which determines the *availability of data* at any given time instant. With this knowledge, EI models, algorithms, and frameworks can be further enhanced to reduce communication costs, energy costs, and end-to-end latency, while maintaining or increasing inference accuracy. This is especially true for online learning that constantly trains models with freshly acquired data, or highly time-sensitive applications, such as remote surgery and autonomous driving. On this regard, the work presented in this book merely scratched the surface of mobile big data for EI, because this field is still in its infancy and faces a number of research challenges that call for future research efforts.

Among these, the biggest challenge is the orchestration among computation devices and communication resources. From the perspective of EI users, the cloud edge continuum is nothing but a giant network of computation nodes interconnected by communication links, in which a fine-grained dynamic cooperation and scheduling mechanism is needed to orchestrate the use of computation and communication resources, under the dynamic availability of training data. There is no simple solution to this problem, as it intertwines with computing hardware design, resource virtualization, scalable/elastic scheduling schemes, and so on. Study on mobile data dynamics provides a different view on resource consumption of data services, leaving traits on data availability that may improve resource allocation and scheduling.

The second challenge comes from heterogeneity of the edge system. Unlike traditional AI that relies on unified high-end equipment in server farms, and neatly compiled datasets, EI faces edge devices (and even end devices) that are diverse in many aspects. In addition to this device heterogeneity, or sometimes also called resource/system heterogeneity, EI also faces data/statistical heterogeneity, as datasets processed by individual devices differ in both size and label distributions, hindering the training and inference in distributed learning schemes. This is further exacerbated by the superposition of dynamic data availability in both time and space domains. Consequently, design of online algorithms and deployment plans will have to work with the knowledge of mobile data with time-varying availability in the edge.

Last but not least, security and privacy remains an unresolved issue for EI. Even though some distributed schemes, such as federated learning, are designed with privacy-preserving objectives at the beginning, it is still possible for attackers to deduce data information from consecutive rounds of parameter updates [8]. Moreover, current security and privacy-preserving measures, such as differential privacy and homomorphic encryption, are quite resource-consuming for the not-so-powerful devices in the edge network [2], while authorization and admission control are also difficult in the open and highly dynamic edge network environment. By adapting the availability of data to the security and privacy needs, we may obtain a finer-grained solution without excessive resource consumption.

As a promising solution to smart healthcare, smart manufacturing, ITS, smart city, and many other applications, EI is gathering tremendous research interests from researchers in various fields. We hope this book provides a fresh angle of data that may inspire further findings toward the provisioning of pervasive EI for users.

References

1. Cao, L.: Decentralized ai: Edge intelligence and smart blockchain, metaverse, web3, and desci. IEEE Intell. Syst. **37**(3), 6–19 (2022). https://doi.org/10.1109/MIS.2022.3181504
2. Duan, S., Wang, D., Ren, J., Lyu, F., Zhang, Y., Wu, H., Shen, X.: Distributed artificial intelligence empowered by end-edge-cloud computing: a survey. IEEE Commun. Surv. Tutor. **25**(1), 591–624 (2023). https://doi.org/10.1109/COMST.2022.3218527
3. Gong, T., Zhu, L., Yu, F.R., Tang, T.: Edge intelligence in intelligent transportation systems: a survey. IEEE Trans. Intell. Trans. Syst. **24**(9), 8919–8944 (2023). https://doi.org/10.1109/TITS.2023.3275741
4. Hayyolalam, V., Aloqaily, M., Özkasap, z., Guizani, M.: Edge intelligence for empowering iot-based healthcare systems. IEEE Wirel. Commun. **28**(3), 6–14 (2021). https://doi.org/10.1109/MWC.001.2000345
5. Yu, D., Xie, Z., Yuan, Y., Chen, S., Qiao, J., Wang, Y., Yu, Y., Zou, Y., Zhang, X.: Trustworthy decentralized collaborative learning for edge intelligence: a survey. High-Confidence Comput. **3**(3), 100150 (2023). https://doi.org/https://doi.org/10.1016/j.hcc.2023.100150
6. Zhang, J., Tao, D.: Empowering things with intelligence: A survey of the progress, challenges, and opportunities in artificial intelligence of things. IEEE Internet Things J. **8**(10), 7789–7817 (2021). https://doi.org/10.1109/JIOT.2020.3039359
7. Zhou, Z., Chen, X., Li, E., Zeng, L., Luo, K., Zhang, J.: Edge intelligence: paving the last mile of artificial intelligence with edge computing. Proc. IEEE **107**(8), 1738–1762 (2019). https://doi.org/10.1109/JPROC.2019.2918951
8. Zhu, L., Liu, Z., Han, S.: Deep leakage from gradients. In: Wallach, H., Larochelle, H., Beygelzimer, A., d'Alché-Buc, F., Fox, E., Garnett, R. (Eds.) Advances in Neural Information Processing Systems, vol. 32. Curran Associates, Inc., Glasgow (2019). https://proceedings.neurips.cc/paper_files/paper/2019/file/60a6c4002cc7b29142def8871531281a-Paper.pdf

Made in the USA
Monee, IL
09 June 2026

52355516R00109